Introduction to Construction Management

Management in the construction industry is a complex task, with team members often undertaking hazardous work, complying with stacks of regulations and legal requirements, and under the constant threat of plans going awry; however, there is no need for all construction management textbooks to be so complicated.

Starting with a general overview of the industry, *Introduction to Construction Management* is the beginner's guide to key concepts, terms, processes and practices associated with modern construction management in the UK. Supported by diagrams, illustrations and case studies, this book explores construction management from a variety of perspectives, including:

- Production management
- Commercial management
- Quality management
- Health and safety management
- Environmental management.

Also incorporated are important industry trends, including sustainability, corporate social responsibility and the advent of BIM. This is the most approachable text available for anyone starting to learn about construction management at HNC/HND, FdSc, or BSc level.

Fred Sherratt is a Senior Lecturer in Construction Management at Anglia Ruskin University. She has over ten years' experience in the construction industry, and worked her way up from the site secretary, through construction planning, to the position of construction manager for a large UK contractor.

Peter Farrell is a Reader in Construction at the University of Bolton and also the programme leader for the MSc in Construction Management. He has delivered undergraduate and postgraduate modules in construction management and commercial management for over ten years. His industry training was in construction planning and quantity surveying and his post-qualification experience was working as a contractor's site manager.

Introduction to Construction Management

Fred Sherratt with Peter Farrell

Routledge
Taylor & Francis Group

LONDON AND NEW YORK

First published 2015
by Routledge
2 Park Square, Milton Park, Abingdon, Oxon OX14 4RN

and by Routledge
711 Third Avenue, New York, NY 10017

Routledge is an imprint of the Taylor & Francis Group, an informa business

British Library Cataloguing in Publication Data
A catalogue record for this book is available from the British Library

Library of Congress Cataloging-in-Publication Data
Sherratt, Fred.
Introduction to construction management / Fred Sherratt with Peter Farrell.
pages cm
Includes bibliographical references and index.
1. Building–Superintendence. 2. Construction industry–Management. I. Farrell, Peter, 1955- II. Title.
TH438.S4826 2015
624.068–dc23
2014010975

ISBN: 978-0-415-70742-8 (pbk)
ISBN: 978-1-315-75522-9 (ebk)

Typeset in Helvetica Neue
by Cenveo Publisher Services

Printed and bound in Great Britain by
TJ International Ltd, Padstow, Cornwall

For Rosie and Zac

Contents

Acknowledgements

The authors would like to thank Simon Sherratt for not being a construction manager, and helping to develop this book as a true introduction to the profession.

They would also like to thank Jennifer Deeney for her time, support and permission to create a health and safety chapter that tells the true story about why health and safety management is so important on construction sites.

They would also like to thank Adept Consulting Engineers Ltd for providing up-to-the-minute BIM data to illustrate how this technology is being used in practice.

Fred would also like to thank all the fantastic construction managers she worked alongside and learnt so much from during her time in industry: Big H, Ramish, Dave, Colin, Steve, Peter M, Dave H, and many more.

Introduction

Modern construction management is a highly complex process: it must ensure that projects and sites are managed efficiently and effectively, that they are on time and on budget, that quality work is carried out, workforce well-being and safety is assured, and environmental impacts are minimised, all while meeting UK legislative requirements.

Management practices must be supported by technical knowledge and good interpersonal skills to bring challenging projects (like that shown in Figure 0.1) to successful completion.

This book introduces you to these management practices, and aims to provide good foundations for you to build upon and help you develop your knowledge and understanding of professional construction management. While the book focuses on the UK industry, many of the concepts and practices discussed here are used worldwide.

Part 1 focuses on the wider construction industry. It explains what the construction industry actually is, and who is part of it. It establishes how industry clients and project teams come together to initiate construction projects, how they are organised and how work is allocated. It looks at the people who work in construction, including

Figure 0.1 A construction site

professions and trades, and how they are supported by various industry bodies and trade unions. This provides the background and the environment in which construction management operates.

Part 2 examines construction management in theory. It looks at what 'management' actually is and the functions of management, what managers actually do, as well as the principles of management and how they relate to construction sites. People management is introduced – vital for a people-focused industry – and communication, motivation, leadership and teamwork are examined in detail.

Part 3 is divided into seven sections which explore different aspects of construction management in practice. Production management is introduced, including the parameters that affect it, as well as the two key aspects of planning and control. These aspects are applied to site management and site layout, time management, cost management, quality management, health and safety management, and environmental management. Each aspect is considered from a practical perspective, and aims to provide you with the skills needed for these different management tasks.

Part 4 considers new directions in construction management, why there are drivers for change within the industry, and what barriers to change need to be overcome. It also introduces more advanced construction management thinking in the form of lean construction and Building Information Modelling (BIM), and how these practices are changing the way we work. The role of the construction manager is also considered, and what new skills may be needed by the construction management professionals of the future.

Throughout the book, construction management is considered in practice – as part of real life work situations with examples and case studies from industry. Rather than simply exploring the theory or ideal outcomes, this book makes these relevant to real life contexts, and also suggests potential solutions to the more common problems of construction management.

At the end of the book, a full glossary of all the common terms, processes and practices of construction management has been compiled for easy reference. These terms are highlighted in blue text throughout the book as they are introduced and become relevant.

At other points in the book, **discussion points** and **exercises** are included.

2.1

Discussion Point

Discussion points are indicated by speech bubbles, asking questions to test your knowledge or to ask you to consider what you have learnt from another perspective. Suggested solutions to all the discussion points may be found at the end of each section; the number of each point may be seen in the bubble.

Exercises require some written work to be carried out. They are alphabetical throughout the text, and the suggested solutions to these exercises may be found in the Appendix at the end of the book.

At the end of each part, **further reading** is suggested alongside the sources cited within the text to enable you to explore further and look up some of the documents or websites mentioned in the book.

1

The construction industry

Introduction

This part of the book aims to 'set the scene' about the construction industry. It will introduce many common concepts, how the industry is organised and how it operates, as well as the professional and industry bodies that have an influence on the construction industry.

It will look at the following subject areas:

- What the 'construction industry' actually is and who is part of it.
- The composition of the industry, and the types and sizes of companies.
- How construction projects come into being.
- How construction projects are organised and how work is allocated.
- Who works in construction, professions and trades.
- Industry trade unions, national agreements and training boards.
- The various industry bodies set up to support it.

To understand construction management and the role of the construction manager, knowledge of the industry and how it operates is needed. This knowledge provides the background to explain why construction managers do many of the things they do, why they face the problems they do, and how they can ensure that they manage to achieve project success in an effective and efficient way.

1.1 What is the 'construction industry'?

The construction industry can be quite complicated to define. There are many different ideas of what the construction industry actually is, and even government departments can spend quite a while trying to explain what they mean at the beginning of their reports about the industry.

This is because the construction industry may be looked at from a number of different perspectives, each of which gives a different idea of what it is.

For example, the construction industry may be considered by type of work or project:

- **New-build** construction projects – built from scratch.
- **Refurbishment** – repairs or work to change the use of a project, for example, turning an old pub into flats.
- **Demolition** work – knocking down part or the whole of an existing project.

These different activities combine to form the construction industry workload.

Civil engineering is another specialist work activity that may be included within the construction industry. This refers to specific types of project such as roads, rail, ports, tunnelling or other types of infrastructure.

Maintenance of the **built environment** also falls within the construction industry remit; for example, redecorating domestic properties or replacing roof tiles are both maintenance activities that will be carried out by people working in the construction industry. This is a growing area in our industry, and has seen the emergence of **facilities management** or FM, which involves both the day-to-day and long-term strategic management of built environment facilities and assets.

The government currently measures construction industry output in terms of just two main categories of work: new work and repair and maintenance (ONS 2014). These categories are further split into housing, infrastructure, industrial and commercial projects, as well as making a distinction between **public** and **private sector** work. Public sector work is funded by the government and the taxpayers, so this may include social housing projects or hospitals, while private sector work is that funded by any other means.

The construction industry may also be considered by the different companies involved in the work; for example:

- Industry **professionals** – the people and companies who design and manage the work, such as architects, engineers and construction managers.
- Industry **trades** – the people and companies who carry out the work using specialist skills, such as plumbers, joiners and bricklayers.
- **Materials** suppliers – the companies who supply the materials used on sites in the construction processes, such as concrete, mortar, nails and sealants. Trade materials are often bought from builders' merchants.
- **Product** manufacturers – the companies who manufacture the products used on the project from their raw materials, such as bricks, timber, windows or roof tiles.
- **Plant** companies – the companies who design, manufacture and supply the tools and plant used on sites to carry out the work, such as cranes, excavators, hammers and fuel bowsers.

There are many other different companies which also contribute to the industry, such as the skip companies that remove and manage the waste from sites, the haulage companies that bring the materials and plant to and from sites, and even the companies that operate the food vans which provide the workforce with their cooked breakfasts!

The UK government, within their Construction 2025 strategy published in July 2013 to set out their goals for UK construction, defined the industry as:

> diverse and its markets broad and varied. Starting with mining, quarrying and forestry, the industry runs all the way from design, product manufacture and construction through to the maintenance of our buildings and infrastructure assets and, at times, into their operation and disposal. [...] Firms in the industry range from world renowned design practices working on some of the most prestigious projects across the globe, to the plumber who turns up on Wednesday afternoon to fix your dripping tap.
>
> (BIS 2013)

This definition also looks at the industry from both a type-of-project perspective and the sort of work involved, as well as from the perspective of all the different companies that work in the industry.

An easy way to bundle all these different definitions up together is to say that, in one way or another, they all relate in some way to the built environment – the buildings and **infrastructure** we live and work in, and use every day.

It is therefore not surprising that construction is a significant industry. In the UK over three million people are employed in the industry, 10 per cent of the total employed, and the industry contributes £90 billion gross value added to the UK economy, which is around 7 per cent of the total. There are over 280,000 companies of various sizes that all work together to form the industry (BIS 2013).

These companies vary considerably in size. Some companies are huge, employing thousands of staff with offices all over the world. Some are smaller in scale, known as small to medium sized enterprises or **SMEs**. SMEs may be either medium-sized and employ fewer than 250 people, or small and employ fewer than 50 people, or micro and employ fewer than 10 people (EC 2013). Also defined as micro SMEs are **sole traders**, individuals who are self-employed and carry out their work alone, for example, an independent architect who produces drawings for small house extensions, or a bricklayer who works alone building them. These micro SMEs make an important contribution to the industry labour force, supporting larger contractors in levelling out peaks and troughs in demand.

In terms of numbers, these different-sized companies are not equally spread out within the industry. There are far more SMEs than large companies, and far more micro SMEs than any other type – 79 per cent of all firms employ fewer than three people and 96 per cent fewer than 13 (ONS 2012).

Because of the large number of smaller companies in the industry, it is often referred to as fragmented.

Construction projects often need to involve many different companies. Even a single house requires lots of different construction activities to complete.

1.1

Discussion Point

Which companies would be involved in building a house?
Are any of these companies reliant on other companies for their work?

All the different companies involved in a project need to be coordinated by someone, the project clearly needs to be organised into elements of work, and this work needs to be assigned to the people with the right skills to do the job. This is where construction management comes in.

1.2 Construction projects

Construction is a project-based industry. Even though lots of the different companies involved in construction such as the materials and product suppliers will work in their own factories, the final output of the industry focuses on construction projects.

The different types of project have already been noted above, and these fall into three main categories:

- New-build
- Refurbishment
- Demolition.

One characteristic of all these three types of project is that they are situated in a particular place, whether that is the land to build upon, or the existing building or structure to remodel or repair, or even knock down.

As a result, our industry is often referred to as transient – the people who work on them have to move around to the different places where the work is needed.

1.2.1 What starts a construction project?

The variety of construction projects means that there are lots of different ways they come into being.

For example, a family may decide to build their own house, or a company that makes shoes may decide it needs a new factory to increase manufacturing, or a successful law firm may decide they need a new skyscraper for their global head office. All of these elements – the family, the shoe company and the law firm – then become what is termed construction **clients** or employers.

It is clients that start construction projects, which are then built for them to meet their specific requirements.

Construction clients start the construction process by deciding that they need a construction project, either to build a brand-new project or to refurbish or demolish something they already own. This will be governed by their own business plans and economic conditions. Why they want the project will influence many other things: how quickly they want it, for what cost, to what levels of quality, and what health, safety and environmental standards are necessary while undergoing construction, as well as how they want the building to perform once it has been completed. Clients may also develop other objectives bespoke to individual projects.

In some cases clients are building **speculative developments** – such as houses, factories, flats or offices – which they can then sell or let as part of their own company business. Some construction companies carry out speculative developments as part of their own portfolio of work, while other speculative developments are undertaken by private clients such as property developers.

One of the largest construction clients is the government, which needs to build schools, hospitals and social housing as well as infrastructure projects such as new railways or roads, water treatment works and ways of producing energy such as power stations or wind farms. The government has a lot of influence on the construction industry, and its policies and development of new legislation can have a significant effect on the industry workload and therefore its success. More about the influence of government may be found in Part 4, which explores the most prominent government reviews, including the latest Construction 2025 Strategy.

But this reliance on clients to start construction projects makes the construction industry very vulnerable to the national economy. If business is booming and the economy is buoyant, then clients are willing to invest in construction projects. But the industry is quickly hit by recession – construction projects are the first to be crossed off the list of things to do and there is usually a long period of good trading before they make it back to market and clients become confident enough to invest again. For this reason the construction industry is often considered to be the 'barometer of the economy' – if construction is doing well then usually the country is doing well too.

1.2.2 How is construction work allocated?

The way construction work is allocated depends on the **procurement route** chosen for the project.

The procurement route is the way a client buys a construction project from the industry. It dictates the way the design, build and operation of the building is allocated to the different companies involved, including the timing of when the companies get involved in the project, how the project is financed, and the way contracts for work are set up. To put it simply, it is the way construction projects are arranged. Procurement routes can be flexible and are not set in stone and clients can arrange their project any way they want, but common procurement routes have developed out of ease of use and good practice.

These common procurement routes have distinct ways and sequences of arranging construction work. If they are often involved in projects, clients may be able to select their own procurement route, such as supermarkets which often commission work to build new stores. Alternatively, one-off clients commissioning their first construction project may need to seek advice from a consultant project manager who is more familiar with how the construction industry works, and so can advise on the most suitable procurement route for the project.

A detailed investigation of construction procurement is beyond the scope of this book, and suggested further reading may be found at the end of Part 1, where you can find out more. However, a brief summary of the most common routes is needed to help explain how construction work is allocated and the potential repercussions of each route.

Procurement routes are all about allocating risk within a project, the idea being that the party best suited to manage the risk is put in charge of it. Key project risks may include unforseen problems within the ground on the proposed project site, bad weather, variations to the design or scope of the project, and inflation or other economic factors. If an unsuitable procurement route is chosen and risks are not managed properly, time and cost overruns may be the result.

There are three main stages to a construction project:

- Design
- Construction
- Operation and maintenance.

The procurement route can influence how these stages are managed, and whether they are kept completely separate or bundled together.

In **traditional procurement** all three elements are kept separate. The client will first consult an **architect** to lead a **design team** (including the architect, **structural engineer** and other specialist designers, depending on the type of project) and produce the design for the project. Once the design is complete, the client then sends the design out to **tender** – the drawings and **specifications** and **Bills of Quantities** for the project are sent out to construction **contractors** to price the work. The tenders or **bids** are then returned and the client selects the best offer. This may simply be the lowest price, but clients may also be looking at other factors than price when they make their decision, such as health and safety record or past experience. These are known as selection criteria.

The traditional procurement route may be illustrated on a timeline, as shown in Figure 1.2.1.

In **design-and-build-procurement**, also known as D&B, the design and construction stages are combined. Clients will first go out to tender on a **design brief** which consists of basic details about the project,

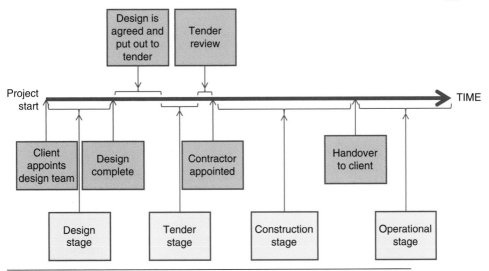

Figure 1.2.1 Traditional procurement timeline

such as location, type of building or any special features it will need. The contractors will then produce not only a price for the project, but also work with their own design teams to produce an outline design as well. The client will then select two or three of those returned to go back out to tender and be developed into more detailed proposals. From these **second-stage tenders** the client then makes their decision to select a contractor for the project on both design and price.

The design-and-build procurement route may be illustrated on a timeline, as shown in Figure 1.2.2.

In **design, build, finance and operate procurement**, more often called DBFO, all three stages are bundled together and another is added – the funding for the project. Usually clients fund projects themselves, but within DBFO projects, a source of funds, such as a bank, pension fund or insurance company, is also involved. Often clients for such projects are government departments or local councils, as using this method of procurement means they are able to obtain new buildings costing millions of pounds for public use, but without needing the capital to fully fund the project themselves. DBFO bidders are usually a consortium made up of a design team, a contractor, a bank and a facilities management company to operate and maintain the project once it has been completed. Again, a **two-stage tendering** process is used, as in D&B procurement. In DBFO projects, clients may pay the consortium a fixed fee each year to effectively lease the facility, and then, after an agreed term – usually around 25 years – ownership of the facility reverts to the client. This is similar to individuals taking a mortgage on their homes; a fixed fee paid each month, including interest, and ownership of the home secured at 25 years. Alternatively, rather than a fixed fee, consortia may be paid on the basis of how much a facility is used (e.g. for prison projects, £x per prisoner per week, or for road projects, £x per vehicle crossing a road sensor). Such projects were previously known as private finance initiatives (PFIs), but

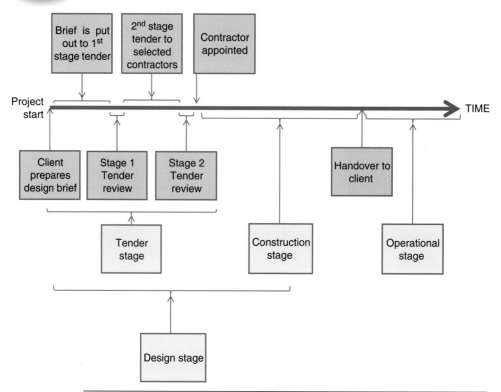

Figure 1.2.2 Design and build procurement timeline

since 2013 this term has been rebranded as PF2, although the DBFO mechanism remains exactly the same.

There are many other construction procurement routes, and elements of each may be combined and used together; for example, in D&B procurement the client may initially develop an outline design with an architect, rather than just use a design brief. Once the contractor has been appointed, the client's architect is then included into the contractor's design team through **novation**.

Each of these procurement routes has its own benefits and potential problems. For example, clients may seek to reduce risk by using D&B procurement, handing the risk for design over to contractors, but this makes any changes they may want to make to the design during the construction phase much more expensive than it would be under a traditional procurement route. As long as clients don't make any changes the risk has been suitably allocated, but if they change their minds, cost overruns of their own making can be the result. D&B gives clients the advantage of single-point responsibility; if a defect becomes apparent during occupation, it may be difficult to diagnose whether it is a design or a workmanship problem. In traditional procurement systems clients may find themselves in the middle of an argument between designers and contractors about where the fault lies. In D&B the responsibility, whatever the cause, clearly lies with the D&B contractor. Since D&B contractors carry more risk, their bid prices to clients may be higher.

These different procurement routes may also be seen at different levels of scale. Although DBFO is restricted to very large projects, traditional and D&B procurement may both be seen on something as small as a domestic extension. Whether the homeowner has an architect draw up the plans which they then pass on to the builder, or whether they approach a small construction company to carry out the work of both the design and build, the principles remain the same.

The above three procurement routes are all linked; traditional procurement developed into D&B, which has recently grown to include DBFO, and the role of the construction manager has developed too. Initially, construction managers were only involved in the construction of projects. D&B procurement then involved them in the design, and construction managers were able to guarantee the **buildability** of projects, ensuring that the design could be efficiently, effectively and safely built. Finally, DBFO made construction managers part of a much larger team, able to influence the design and construction alongside those funding and maintaining the finished project.

Therefore construction managers need to be able to fit into a number of roles and carry out work from a number of different perspectives, depending on the procurement route of the project. They may be operating alone, taking instructions from an architect, or working as part of an integrated project delivery team.

Discussion Point

1.2

If you were in a position to have a new house built for yourself, how would you procure it?
Who would you select to design it?
How would you establish estimates of cost?
Who would you want to take the risk for uncertainty about the ground conditions?
If the completion date of the house was delayed by bad weather, who would you want to pay for the cost of the delays?

1.2.3 How are construction projects organised?

A new term was introduced above, namely the construction contractor, who tenders for construction projects.

The reason for this term is the way construction projects are organised.

Whatever procurement route is used, once clients have received the tenders for their work and appointed a contractor, a formal **contract** is set up between them for the work to be carried out. The procurement route can influence the type and terms of the contract used, but essentially a binding legal agreement is made in the form of a contract. Common construction contracts include the Joint Contracts Tribunal (JCT) suite of contracts and the New Engineering Contracts (NECs).

However, on many construction projects the work is very complicated and requires many different specialist skills. Just one company could not carry them all out and so the contractor needs to develop a construction team who can provide the necessary trade skills to complete the work and passes all or some of the work on to other specialist companies. They also do this through the use of legal contracts – the **main contractor** is the company that won the work, while those working for the main contractor are termed **subcontractors**.

These subcontractors may also allocate part of their work to other specialists as well, who then become sub-subcontractors in turn. Both the main contractor and subcontractors will also have their own materials, product and plant suppliers. These layers may be termed tier 1, tier 2, tier 3, etc., the further they are positioned from the client. The contractual links through a project can therefore become quite complicated, and all these different companies linked together are what is termed the project **supply chain**. An example supply chain is shown in Figure 1.2.3.

Construction work is often allocated to the supply chain through the use of **work packages**. Work packages may be aligned to the specialist trade needed to carry out the work, such as the brickwork for a project. This would be allocated as a brickwork package to a brickwork subcontractor.

Alternatively, larger sections of the project may be allocated as a bundle; an example would be the **envelope works**. This would include the brickwork, **cladding**, **curtain walling** and windows – all the elements that make up the outer skin (envelope) of the building. The package subcontractor may then subcontract out the elements of this work they cannot do themselves (e.g. the brickwork and cladding), and keep the curtain walling and windows in-house. The envelope package could also have been allocated, with some design work included,

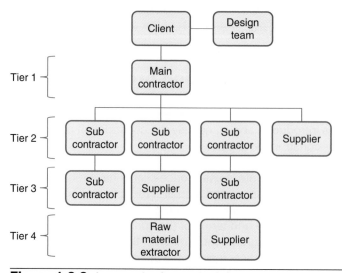

Figure 1.2.3 An example of a supply chain

through D&B procurement, to the envelope subcontractor who either carries out the design themselves or subcontracts it to a specialist. The risk for the design and construction of these four envelope elements has been passed to the company who can best manage it and deliver four different elements of the project as one integrated package through their own supply chain. This also means that the main contractor has a single point of responsibility should defects or maintenance issues arise with the envelope once the works have been completed.

Construction industry supply chains are often complex; they are very fragmented and can often overlap. Supply chains include all parties and companies that contribute to the construction process and its end products, and explain how SMEs and even micro SMEs can work on large projects for large contractors. Supply chains can be both beneficial and also problematic for construction managers; this is discussed in more detail in Part 4.

1.2.4 Who forms the construction project team?

So far, we know that the project team consists of the following:

- The client – who starts the construction project, and pays the bills.
- The design team – led by the architect, this may also include other professional designers such as structural or civil engineers, mechanical and electrical service designers or acoustic specialists.
- The main contractor – who wins the project and manages the build, including the role of the construction manager. Some main contractors may carry out some of the work themselves; for example, they may have their own joiners or bricklayers, while others will subcontract all the work.

The design team and main contractor are often termed the **professional team** – the people who design and manage the work.

- Subcontractors and suppliers – who provide the trade skills for the project, and carry out the work packages allocated by the main contractor.

The subcontractors and suppliers are often termed the **operational team** – the people who actually do the work.

One construction industry profession is missing from this list – the **quantity surveyor** or QS, who manages the cash flow, costs and contracts for the project. QSs will work for either clients or contractors.

The different members of this team may become involved in construction projects at different times depending on the procurement route, but whenever they are brought on board they must work together to achieve project success. Whichever procurement route is followed, there needs to be clear communication and collaborative working practices to ensure that information is available when needed, accurate estimates of time and money are delivered, and that the team members are able to meet the client's needs efficiently and effectively.

1.3 Construction people

Working in the construction industry can be very rewarding – it is very satisfying to walk past a fantastic building and be able to say, 'I helped build that!', or even better to exaggerate and say, 'I built that!'

But the construction industry also faces challenges owing to a poor image which affects its ability to recruit and retain a strong and talented workforce (BIS 2013).

Discussion Point

Think about the image of the construction industry.
What words would you use to describe it?

Problems are caused by the lack of knowledge young people have about the opportunities available in the industry, as well as its poor health and safety record. Personal involvement in the industry can also give people a bad impression; the domestic repair and maintenance market receives more complaints than any other from the general public, about so-called 'cowboy builders' who don't complete on time and may charge too much for poor-quality work. The industry also needs to improve its levels of equality and diversity; less than 14 per cent of the workforce is female and only 2 per cent are from ethnic minority groups (BIS 2013).

These are all problems which the industry, and the UK government in its Construction 2025 Strategy, are trying to resolve.

Various initiatives such as the Construction Industry Training Board's (CITB) *Construct Your Future* campaign are attempting to tackle the poor image and industry recruitment issues, by raising the awareness of young people about the careers available in construction both in the professions and through trade apprenticeships (CITB 2013a). It is essential that more young people enter the industry, as the current workforce is ageing; 500,000 people will retire from construction between 2014 and 2020, and the industry will needs to recruit 165,000 people every year just to maintain current numbers of employees (CIC 2013).

The Health and Safety Executive (HSE) is helping the industry to tackle its poor health and safety record, and in 2012/2013 the lowest figures for industry fatalities were recorded in the UK to date. Ongoing campaigns include the Safer Sites series run by the HSE, which aims to support continued improvements in health and safety standards in the industry (HSE 2013a).

Domestic repair and maintenance companies can register with the TrustMark Scheme, which provides assurance that any work will be carried out in accordance with government-endorsed standards. To become TrustMarked, a company must be inspected and approved by one of the scheme's operating bodies. Of all the jobs carried out by TrustMark firms, there was only a problem with 0.01 per cent of them, far less than the complaints received about non-registered firms (BIS 2013). TrustMark is able to reassure the general public that construction work can be carried out competently and efficiently, and is doing its best to improve the image of the industry.

The problems with equality and diversity can be traced back to all of the above concerns, and the need to attract an equal and diverse workforce into the industry in the first place. The vast majority of the industry is white and male, but companies are actively seeking to diversify by improving opportunities for employment and awareness of the potential for both professional and trade roles within the industry.

While construction may face challenges in terms of its people, it is still a great place to work. There are still well-established and acknowledged professions and trades, supported by their professional bodies and unions, which are proud to work on construction projects every day.

1.3.1 Construction professions

Within the project team, the design team (architects and engineers), quantity surveyors, consultants and construction managers are also known as the professional team.

Professionals are identified through membership of their own **professional bodies**. A professional body is an organisation that ensures professional standards are maintained. Membership of a professional body can give an individual chartered status.

Different construction industry professions and different professional bodies have led to a level of fragmentation within the professional team. It has been suggested that the separation of the design, cost and construction elements of construction projects has partly been to blame for the industry problems of time, cost, quality, health and safety, and environmental issues. If professionals are only focused on their one aspect of the project, this can cause problems in other areas, such as a lack of buildability or budget overspends.

However, professional teams can work very successfully together, and many successful construction projects demonstrate this. Looking forward, the need for teamwork and collaboration between the professions is growing, and the use of **building information modelling** (BIM) will hopefully accelerate the process of producing integrated teams for project delivery. There is more information about BIM in Part 4.

Architects and the RIBA

Architects are the designers of the project; in the traditional procurement route they lead the design team and also have a managerial role. In the UK they are the only construction industry profession that must be registered with the Architect's Registration Board, which provides statutory regulation of architects. Architects have formal qualification systems through degree programmes accredited by their professional body, namely the Royal Institute of British Architects, or RIBA.

agreeing monthly **valuations**, and seeking out best value for the client where possible.

The role of the main contractor or subcontractor's QSs involves pricing and payment for the work, the negotiation of the final price, and the agreement of the contractual arrangements for the project. QSs will also be keen to seek out value from subcontractors and suppliers, and will closely monitor the rate of spend and cash flow on the project.

Quantity surveyors are supported by the Royal Institution of Chartered Surveyors (RICS), which also includes land surveyors and valuation surveyors for existing properties. The RICS was established in 1868, and aims to:

> regulate and promote the profession, maintain the highest educational and professional standards, protect clients and consumers through a strict code of ethics, and provide impartial advice, analysis and guidance.
>
> (RICS 2013)

Construction managers and the CIOB

Construction managers include all those involved with the management of construction work, and so construction managers can work for the main contractor, subcontractors and even clients in the role of their consultant project manager.

The professional body is the Chartered Institute of Building (CIOB), which started out as the Builder's Society in 1834, and:

> is at the heart of a management career in construction. We are the world's largest and most influential professional body for construction management and leadership. We have a Royal Charter to provide the science and practice of building and construction for the benefit of society [...] the development, conservation and improvement of the built environment.
>
> (CIOB 2013)

Other professional bodies

There are also many other professional bodies within the construction industry, each representing a different skilled profession and providing chartered status for its members.

For example, mechanical and electrical installations within a project can comprise up to 50 per cent of the total project costs and can be highly complex in their design and installation; professionals working in this area are represented by the Chartered Institution of Building Services Engineers (CIBSE).

Becoming a chartered professional provides a measure of an individual's knowledge, skills and integrity, and is an important stage for anyone seeking a professional career in the construction industry.

Becoming chartered

To become chartered, individuals need to follow the specific requirements for academic qualification and practical experience and training set out by their professional body. Becoming chartered is not simply about passing exams or gaining a degree; it is also closely linked to professional practice and experience.

All professional bodies hold a code of ethical conduct that their members must adhere to in their professional practice. **Continuous professional development** (CPD) is also a requirement for chartered status, and members need to demonstrate that they are continuing to learn and keep up to date with the latest industry developments.

Within all professional bodies there are various levels of membership, and this often includes free membership for students. It is a good first step in your professional career to become a student member of your professional body and to attend CPD and networking events in preparation for becoming a full member in the future.

1.3.2 **Construction trades**

Construction trades are the people who actually carry out the work. There are a huge number of different construction trades, all of which require different skills, knowledge and training. People who work as construction trades are also referred to as **operatives** – again, this term simply means the people who practically carry out the work on sites.

What defines a construction trade?

There is no definitive list, but the list of trades for which a specific Construction Skills Certification Scheme card may be issued numbers 232, and runs from access floorers to wood preservers/damp-proofers.

Common construction trades include **groundworkers**, bricklayers, electricians and plumbers. There are also non-specific trades. For example, **general operatives** carry out general unskilled works on the site, such as keeping the site tidy or moving materials to the workface, although they may occasionally also carry out skilled works, such as basic concreting. 'Gen-ops', or **labourers** as they are also sometimes termed, undertake whatever basic and unskilled work is needed on the project as work progresses.

Many trades have their own specific routes for training and qualification.

City and Guilds qualifications enable young people to enter a trade through an **apprenticeship** or by gaining recognised National Vocational Qualifications (NVQs). Qualification programmes involve practical skill-based learning alongside academic programmes, which include aspects such as health and safety management on sites.

Apprenticeships involve people working and learning on the job, as well as attending college for academic support, so students can earn while they learn.

National Occupational Standards for the different trades are maintained by the **Construction Industry Training Board** (CITB), which sets out the skills and knowledge needed for specific trades at specific levels. For example, for bricklayers, the Standards for Trowel Occupations (Construction) for Level 1 require students to 'prepare and mix concrete and mortars' and 'lay bricks and blocks to line', among other skills. Level 3 requires students to 'erect complex masonry structures' and 'confirm work activities and resources for the work' as part of the mandatory requirements.

The CITB provides training through the National Construction College and produces industry-relevant publications, and helps administer the industry Construction Skills Certification Scheme (CSCS) for on-site health and safety as well as associated specialist schemes for scaffolders and plant operatives. The CITB states:

> It's our job to work with industry to encourage training, which helps build a safe, professional and fully qualified workforce.
>
> (CITB 2013b)

The CITB is funded by an industry levy, meaning that construction industry companies invest in the training and development of their own workforce. The levy is payable by all companies with a wage bill of £80,000 or more.

The CITB also works in partnership with the Construction Industry Council (see Section 1.4 for more information on the CIC) and CITB-Construction Skills Northern Ireland to form the Sector Skills Council for Construction – known as ConstructionSkills.

Find out more about the CITB on its website.

How do trades work on sites?

Many construction trades will work collaboratively in **gangs** to produce the most efficient output for their trade.

For example, bricklayers will often work in what is known as a 2+1 gang: two bricklayers working with one labourer as a team. The bricklayers will just lay bricks, while the labourer makes sure that enough bricks and mortar are available for them to work continuously, and also keeps the work area clean, tidy and safe. Often gangs are paid for the work they produce as a team or are tasked with an area of work to complete, and share payment among them.

In larger gangs or teams, a **supervisor** or **foreman** will usually be nominated to help manage work for the team. Supervisors will check the project information and drawings, make sure the right materials are available and assign work to the different members of the gang. Supervisors sometimes carry out some of the work themselves.

It is the supervisors and foremen who provide the link between the construction manager and the site trades, and they should be able to pass information effectively and efficiently between the two. Often they will be involved in the short-term planning of the project with the construction manager, and agree milestones for areas of work to be completed to hand over to following trades.

Supervisors and foremen often receive training for their role, but sometimes their appointment is simply based on experience and knowledge. In some cases formal training may be required, such as health and safety training for supervisors provided by the Site Supervisors Safety Training Scheme (the SSSTS) run by the CITB, which may be a requirement for supervisors working on large projects.

Construction trade unions

While the construction professions are represented by their professional bodies, construction trades are represented by their **unions**. The construction industry has generally had a good industrial relations record, but unions are still able to provide support in cases of discrimination or unfair dismissal.

Three unions are most prominent in the construction industry: the Union of Construction, Allied Trades and Technicians (UCATT), the General, Municipal, Boilermakers and Allied Trades Union (GMB), and the Transport and General Workers Union (T&G).

UCATT is the only UK trade union dedicated to construction workers. It supports its members in securing fair rates of pay and conditions. It is trying to increase levels of direct employment in the industry, to ensure that workers are able to access basic employment rights such as holiday and sick pay. UCATT was also key in the drive to develop a fully registered workforce through the CSCS scheme, to improve safety on sites.

The industry unions were instrumental in the production of various industry **Working Rule Agreements** (WRAs) which set down the general conditions for employment within the construction industry and are negotiated with employers' associations. There are several different WRAs in operation, including the *Construction Industry Joint Council Working Rule Agreement for the Construction Industry* and the *Building and Allied Trades Joint Industrial Council Constitution and Working Rule Agreement*. There are also WRAs specifically for civil engineering, electrical and heating and ventilation works.

The WRAs set down key elements about work and working conditions, including the following:

- Rates of pay
- Working hours
- Extra payments relating to skills and working conditions
- Overtime payments
- Holidays

- Travel allowances
- Health and welfare conditions.

Although some employers opt out of WRAs, it is the role of the construction manager to ensure that any agreed WRA conditions are met on their site, and they should also ensure that fair employment of operatives is being carried out by any subcontractors working on their projects.

1.4 Construction bodies

Over time, and often in response to various industry reviews undertaken by different governments (more information about these reviews may be found in Part 4), many different industry bodies have been established. These bodies aim to address issues, drive change and bring the various industry contributors, such as clients, professionals, contractors, product manufacturers and other suppliers, together. A key development was the creation in 2008 of the role of the Chief Construction Adviser to the UK government, an acknowledgement of the significant role the government has as an industry client. This role was created to work with the various construction bodies, as well as champion the construction industry in the UK.

Three bodies are introduced below; these are some of the most prominent within the industry and all have been established to support or give voice to the construction industry, although some do overlap in their remit and how they operate.

However, all industry bodies are aiming to improve industry performance within their own specific scope. These organisations are an excellent source of information and good practice, and an effective way to keep up to date with the latest industry developments.

1.4.1 Constructing Excellence

Formed in 2003, Constructing Excellence is the result of bringing together many established cross-industry bodies to create a single voice for improvement in the construction industry. It has been charged with driving change to improve industry performance. Constructing Excellence includes members from across the sector and the sector supply chains, and is member-led.

It undertakes research and demonstration projects to gather evidence of what works for the industry and to show the benefits of the adoption of change. You can see some of the demonstration projects on its website. It communicates through various industry networks and also provides guidance and training on how to implement change. It also hosts membership forums to support these activities, and brings together client and industry groups to drive change forward.

More about Constructing Excellence may be found on its website.

1.4.2 Strategic Forum for Construction

The Strategic Forum is the main interface between the government and industry. It brings together the main industry bodies and aims to drive construction towards the goal of delivering maximum value for all its clients through the consistent delivery of quality projects.

The Strategic Forum is made up of the following:

- Construction clients, represented by the Construction Clients Group.
- Professional bodies, represented by the Construction Industry Council (CIC).
- Contractors, represented by the UK Contractors Group (UKCG) and the Construction Alliance.
- Specialist contractors, represented by the National Specialist Contractors Council and Specialist Engineering Contractors Group.
- Product suppliers, represented by the Construction Products Association.
- Construction site workers, represented by UCATT on behalf of the unions.

This list of members clearly shows the large number of different groups and their specific interests within the UK construction industry.

The Strategic Forum aims to promote six key areas contained within its construction commitments:

- Procurement and integration
- Commitment to people
- Client leadership
- Sustainability
- Design quality
- Health and safety.

The Forum produces reports and toolkits to help the industry focus on these aims; for example, through the *Respect for People* initiative the Forum also helps focus on health and safety in the industry to drive improved performance and bring about business gains.

More information about the Strategic Forum for Construction may be found on its website.

1.4.3 Construction Industry Council

The Construction Industry Council (CIC) is the representative forum for industry professional bodies, research organisations and specialist business associations. Through these organisations, the CIC has a collective membership of 500,000 professionals and 25,000 firms of construction consultants, and seeks to represent their interests through a single voice.

It aims to give leadership to the UK industry through participation in many other industry initiatives and programmes such as the Strategic Forum for Construction, as well as promote quality and sustainability within the built environment. More information about the Construction Industry Council may be found on its website.

1.5 The construction industry: summary

Part 1 of the book has set the scene for the UK construction industry and provides the background for modern construction management.

Construction projects are initiated by clients through one of many different procurement routes that enable them to 'buy' their construction project from the industry. Work is organised with both professional teams including designers, quantity surveyors and construction managers, and operational teams made up of the construction trades who actually carry out the work on sites. These ways of working have created a fragmented industry, with long supply chains of subcontractors and suppliers, and a transient workforce moving from project to project.

It is often argued that this structure and organisation has led to many problems; for example, projects that go over time and budget, poor-quality work and bad health, safety and environmental records. Over the past decades there have been developments in how we think about construction as an industry and how we organise construction projects. It has been suggested that the industry must work together more to provide best value to construction clients, and how its role and the role of those working within it should change to support its future.

Further reading

The following textbook takes a closer look at the UK construction industry and how it works:

Morton, R. and Ross, A. (2008) *Construction UK – Introduction to the Industry*, 2nd edn, Blackwell, Oxford.

You can find out more about tendering, procurement and contracts in the following textbook:

Harris, F. and McCaffer, R. (2013) *Modern Construction Management*, 7th edn, Wiley-Blackwell, Chichester.

Find out more about the professional bodies on their websites:

CIAT at www.ciat.org.uk
CIOB at www.ciob.org.uk
ICE at www.ice.org.uk
RIBA at www.architecture.com
RICS at www.rics.org

More information about the RIBA plan of work may be found on its website:

www.architecture.com/TheRIBA/AboutUs/Professionalsupport/
RIBAOutlinePlanofWork2013.aspx

More about the Construction Industry Training Board may be found on its website:

www.citb.co.uk

The industry bodies also have useful websites:

Constructing Excellence at www.constructingexcellence.org.uk
The Strategic Forum for Construction at www.strategicforum.org.uk
The Construction Industry Council at www.cic.org.uk

Discussion point comments

Discussion Point

Which companies would be involved in building a house?
Are any of these companies reliant on other companies for their work?

For a domestic project the companies providing the following trade skills may be needed:

- groundworkers (foundations and drainage and GF slab)
- bricklayers
- joiners
- roofers
- dryliners (partition walls)
- plumbers
- gas fitters
- electricians
- plasterers
- window fitters
- painters and decorators
- floor layers
- kitchen fitters.

Many of these trades need to work together. For example, the bricklayers, joiners and roofers need to work together to ensure that the gable ends of a house are constructed correctly. Plumbers and electricians will have to work with the kitchen fitters to ensure that everything is connected up correctly.

It is possible that a company may be able to provide more than one trade; for example, a small services company will manage both electricians and plumbers within the same organisation.

Discussion Point

If you were in a position to have a new house built for yourself, how would you procure it?
Who would you select to design it?
How would you establish estimates of cost?
Who would you want to take the risk for uncertainty about the ground conditions?
If the completion date of the house was delayed by bad weather, who would you want to pay for the cost of the delays?

As an occasional user of the construction industry, you may wish to minimise your risks; this is more likely to be the case if your budget is limited. However, if you pass risks on to another party, that party will often increase its price. Thus you will pay £x if you take the risk, or £x plus if another party takes the risk. The problem is that if things go wrong and you take the risk, you may end up paying £x plus-plus. Using some figures to illustrate, your options are £100k, £110k or even £120k. Alternatively, £105k and the contractor takes the risk.

In such cases, it may often be the case that individuals will approach architects to produce a design, and process that design through the planning and building regulation stages. Architects will be happy to provide some indicative costs and to conduct a tendering process with bidding contractors on your behalf, although architects may obtain some help or advice from quantity surveyors. Normally, contractors would expect you as the client to take the risk of the ground, and ask that you allow extra time to complete if the weather conditions are exceptionally inclement. All this is analogous to the traditional procurement method. You will be left with split responsibility for design and build, and some uncertainties about whether a bid price of £100k will be exceeded. You may need to push contractors a little harder if you wish them to take responsibility for the design and the ground, and be prepared to pay an extra sum of money for this. A comprehensive ground investigation report will mean that contractors will have at least some information about the ground, and they will not feel the inclination to add large sums to their bids to cover for ground risks. It may be reasonable that you allow extra time to complete in the event of exceptionally inclement weather; unless of course your personal circumstances dictate that a particular date must be met. In that case, expect the contractor to add a sum of money to its bid to cover the risk involved.

However, as a construction manager, you are highly skilled, have excellent knowledge and are highly motivated to minimise the impact of any problems that may arise. Therefore you take all the risks. You manage the construction so that you are not paying the salaries of a contractor's staff, and not paying a contractor profit. You employ subcontractors to complete various packages of work. Perhaps you employ a self-employed general operative to keep the site clean, safe and to complete any preliminary-type tasks. Perhaps you can complete the project for £90k or less.

Discussion Point

 1.3

Think about the image of the construction industry.
What words would you use to describe it?

The industry can be an exciting, challenging and dynamic place to work, which can provide those who work in it with a great sense of

achievement. There are also opportunities to travel, both nationally and internationally, and it is a good industry for entrepreneurs and people wanting to work for themselves.

Sadly, the construction industry often has a poor image within the wider general public perspective. It may be seen as dirty and dangerous, and the work conditions can be cold and wet or hot and dusty depending on the time of year. It is often seen as a macho industry, dominated by men and supported by the image of the big, strong construction worker. Unfortunately this has led to health problems, with many workers feeling that they have to fit into this stereotype and causing themselves harm by lifting too much or not protecting themselves properly from harmful materials.

The industry is often seen as just its sites and the trades, which can mean that the professional and management side is not fully appreciated.

PART **2**

Construction management in theory

Introduction

This part of the book is all about the theories behind the
'**management**' in **construction management**.

A theory is a way of thinking about things in an abstract and objective
way; for example, to explore why things are done the way they are or
why people act the way they do in certain situations.

This part aims to develop theoretical understandings of what
management is, which may then be used as a foundation to help
develop a deeper understanding of construction management in
practice.

It will look at the following subject areas:

- What is 'management'?
- The functions of management, and what managers actually do.
- What are the principles of management and how they relate to
  construction sites.
- The social nature of construction work and how this affects
  construction management.
- How construction managers can motivate people at work.
- What is leadership, and how it is different to management.
- Teamwork, and how it is vital at all levels of construction industry
  operations.
- What is construction management?

There are thousands of books written on the subject of management,
and many theories of management have been developed by academics
and practitioners over the years. A lot of work has also been carried
out around construction management, looking at how management
operates specifically within our industry.

Part 2 sets out to give you an introduction to management functions
and principles, and explores how they fit in with modern construction
management and construction work on a day-to-day basis.

2.1 What is management?

Unfortunately there is not a simple answer to this question.

Despite the vast amount of research carried out about management in the many different businesses and industries around the world, there is no one accepted answer. There are indeed hundreds of definitions of 'management'.

Discussion Point

What do you think management is?
Can you explain it in just a few words?

One of the simplest definitions was put forward by Calvert *et al.* (1995), who said that:

> Management is the conduct and control of organised human activity.

From this definition we can see that Calvert suggests there are two key aspects to management: conduct and control.

Control refers to the practical aspects of management – what managers actually do on a day-to-day basis. This is often referred to as the 'science' of management, the defined tasks to be carried out for effective and efficient management. These are also known as the **functions** of management.

Conduct refers to the people aspects of management – how managers work with the people under their control on a day-to-day basis. This is often referred to as the 'art' of management, because people are not simple or straightforward and a scientific 'one-size-fits-all' approach just would not work. For the functions to work effectively, 14 **principles** of management have been identified in order to harmonise conduct with control. Other factors such as teamwork, motivation and leadership are also important in **managing people**.

The following three sections introduce the functions of management and the principles by which they should be carried out, and begin to explore the more important aspects of managing people in greater detail.

2.2 Management functions: what do managers do?

In their definitions of construction management, both Fryer *et al.* (2004) and Griffith and Watson (2004) make reference to the *functions* of

management, which were initially put forward in Victorian times by a man called Henri Fayol.

Fayol was a mining engineer who became one of the most influential figures in the field of modern management. Fayol's management functions are still used and often referred to today, and form the basis of most definitions of management, including construction management.

These functions may be used to describe what managers actually *do* in their daily jobs: how they get the job done. The management functions are illustrated in Figure 2.2.1.

Each of these functions will be be examined in more detail from the perspective of construction managers.

2.2.1 Forecasting

Forecasting is looking ahead into the future to predict possible trends or occurrences which are likely to influence the working situation. Forecasting for a construction company will be carried out at a variety of levels, focused on different horizons into the future, as is shown in Figure 2.2.2.

Discussion Point

What might directors of large construction companies be trying to forecast?
What might construction managers be trying to forecast?

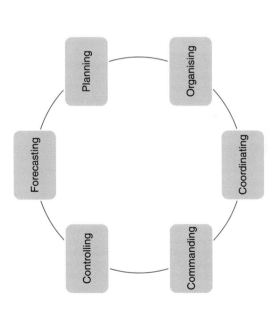

Figure 2.2.1 Fayol's Management Functions

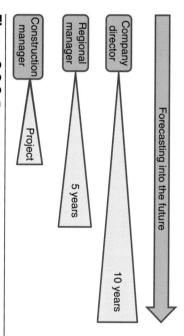

Figure 2.2.2 Forecasting into the future

2.2.2 Planning

All managers plan; this is the development of the 'campaign of action' designed to achieve specific objectives. Again, there are different levels of planning which occur over different time-scales, depending on the level of management involved.

Strategic planning is long term and covers a period of years. This is undertaken by senior management teams within companies in response to the forecasted trends, to ensure that they are able to meet the future successfully.

Planning by construction managers is likely to be for the duration of specific projects, but will also involve more detailed **operational** planning at the site level.

Planning is a key aspect of construction management. Larger contractors often have dedicated planners within their construction management teams; for example, pre-tender planners have a role in estimating teams trying to win the work, while project planners work within site teams and continue to plan as projects progress on sites.

Discussion Point

Why does construction work need to be planned? What elements would a construction project plan need to take into account?

2.3

Planning documentation can take the form of a company policy, a five- or ten-year strategic plan for the company, or the site **programme** which is used to plan the work needed on a weekly or daily basis to meet the final project completion date. We will look at this operational level of planning in much more detail in Part 3.

In many companies, the processes of forecasting and planning are merged and undertaken as one management task. Forecasting is the process of looking into the future while planning the process of making decisions and developing the plan based on that forecast.

2.2.3 Organising and coordinating

This involves putting the plans made during the planning stage – which were themselves developed from the forecasting stage – into action. Organising aligns closely with coordinating when considered from the construction management perspective.

Construction managers are concerned with the allocation of tasks to certain people or teams, ensuring that the correct resources (**labour**, **plant** and **materials**) are on site and allocating them correctly to enable people to carry out their given tasks, and setting deadlines but also ensuring that people are able to meet them. This also involves coordinating all the individual tasks to make sure that everyone can work harmoniously towards the project goal of successful completion.

For example, in the construction of a domestic property it would not be effective for all the internal **trades**, namely the electricians, plumbers, joiners and plasterers, to be in the same space at the same time – no one would have the room to do their work properly or safely. Therefore the construction manager must organise and coordinate work to ensure that the tasks are completed in the correct order and to the correct standards, in order to meet the required objectives.

Discussion Point

How does the use of subcontractors influence the allocation of tasks at site level?
What effect does this have on the coordination role of the construction manager?

2.2.4 Commanding

What began as commanding in Victorian times has now developed into *motivating*. The use of motivation is now considered a far more effective strategy than commanding to encourage employees to produce their maximum outputs. Encouragement is believed to achieve better results than pressure and punishments.

Construction managers must therefore create the correct environment for employees to *want* to give their best efforts to their work, and to align their own personal goals with those of the companies or projects. Motivation is a key aspect of construction management, and the way work is organised on construction projects can make this function very complicated to implement in practice. Motivation is examined in more detail in Section 2.4.2.

2.2.5 Controlling

Controlling involves the monitoring of both the forecasting and planning activities, and checks that the action taken in organising and coordinating has achieved the planned objectives. It enables construction managers to ensure that their projects are progressing as planned and, if not, corrective action can be taken.

The planning documentation, the programme for the work, forms the basis for controlling. Without programmes, managers cannot control work. But if managers do not control performance then programmes are of no value. Planning and controlling are interdependent, and are supported by **feedback** provided through the organising and coordinating functions.

In order to make sure that the programme is up to date and the future plan is accurate for the next stages of the project, feedback is needed to reflect what work has been done so far. This feedback is also needed to check that the initial plan was effective. If a task was planned to take a week and it took four weeks, something may have been wrong with the programme to start with!

Areas requiring control are time, cost, quality, health, safety and the environment:

- Time is controlled through the programme or plan of work, and performance is measured against the plan.
- Quality is measured against specifications and regulations and is often controlled through **inspections**, although the need for inspections can be seen as a failure to get it 'right first time'.
- Cost is measured against project budgets, quotations and **Bills of Quantities (BoQs)** and is controlled through close **monitoring**, usually by quantity surveying teams.
- **Health** and **safety** are measured against company policies, legislation and other project-specific objectives, as well as through the numbers of accidents, incidents and near misses. They are controlled through method statements and risk assessments, constant monitoring and regular inspections.
- **Environmental** control is measured against company policy and legislation, and controlled through method statements, risk assessments, inspections and ongoing monitoring.

Controlling functions should be performed on a regular basis, and corrective action taken as necessary. Controlling therefore becomes a cyclical process to further check that the corrective action has met requirements and put the work back on course, as shown in Figure 2.2.3.

Many companies refer to the controlling function as 'review' or 'monitoring' activities.

Discussion Point

What corrective action could be taken if performance is not meeting the original plan?

What corrective action could be taken if cost is not meeting the original plan?

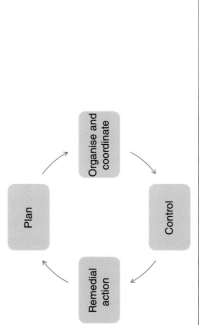

Figure 2.2.3 Fayol's Functions adjusted to show remedial action

2.2.6 How important are these functions for a construction manager?

Griffith and Watson (2004: 33) report the findings of a study that asked 50 construction managers which of Fayol's management functions were most important to them in their work.

This study found that planning was believed to be the most important function, followed in order of importance by organising and coordinating, controlling, and motivating. Forecasting was the least important.

It is not surprising that planning came top of the list, as without clear and effective plans the other functions cannot be carried out effectively. Forecasting may be seen as the least important of the functions because many construction managers will embed this function alongside planning due to the project-based nature of their work.

These functions may often merge to become an overall managerial process, and are not necessarily linear or undertaken separately. As we have seen, some functions overlap and some combine, and in fact the overall process may be seen to be cyclical, with the feedback from the controlling element informing future forecasting.

Part 3 of this book examines how these functions of planning, organising, coordinating and controlling are actually put into practice by construction managers.

2.3 Principles of management: how do managers do it?

The principles of management were also set out by Henri Fayol, who developed them through reflection and consideration of how best to implement his management functions within the social environment of work.

The principles aim to set out fundamental 'truths' about management, and clearly list what is required for *effective* and *efficient* management of people at work. Again, the principles apply to all fields of management, including construction management.

The principles are shown in Table 2.3.1.

These management principles will be examined through a scenario: the construction of the foundations on a new housing development of 20 plots. Wide strip foundations with one layer of mesh reinforcement are to be used in the excavation, as is shown in Figure 2.3.1.

Using this scenario as our basis, we will now look in more detail at the first principle of management:

The division of work: Work should be divided so that attention is given to small portions of a bigger task and employees should be allocated portions they are skilled in. This specialisation of work increases skill and efficiency.

This principle states that work should be divided into smaller tasks, using skilled employees who will become more skilled the more often they carry out the same tasks.

This is clear enough in terms of management activity, but to understand *why* the principle is of use it is often a good idea to consider it from the other way round: to think about what could go *wrong* if work isn't carried out in accordance with this principle. In this case, there could be problems with the following:

- **Inefficient work** – if the work is not divided into smaller tasks there could be poor sequencing and timing in constructing the 20 foundations, as the workforce may try to complete one house before starting on the next, wasting labour and plant time as they wait for preceding tasks to complete. The work can also take longer if unskilled workers are used, due to their lack of familiarity and ability to complete the tasks efficiently.

- **Poor-quality work** – if workers are not skilled in their allocated tasks poor-quality work may result. Inaccurate setting out could result in

Table 2.3.1 Fayol's management principles

Division of work	Centralisation
Authority and responsibility	Scalar chain
Discipline	Order
Unity of command	Equity
Unity of direction	Stability
Subordination of individual interests to the general interest	Initiative
Remuneration	Esprit de corps

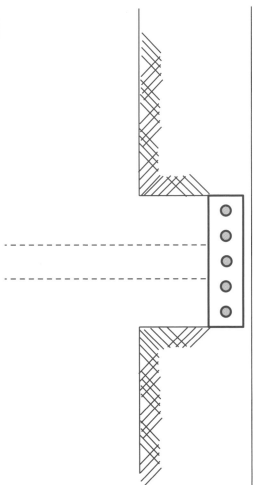

Figure 2.3.1 Section through a wide strip foundation

the foundations being cast in the wrong locations. Poor concrete work, using badly laid concrete with incorrect reinforcement cover, or even the wrong strength of concrete, could all lead to defects later on.

- **Inconsistent work** – if work is not carried out by the same operatives on the different plots this may lead to inconsistent defects later on.

- **Unsafe work** – if everyone is trying to work on one plot at the same time this could result in people working too close to the excavator and being put at risk.

All of the above will cause problems for construction managers in terms of potential accidents on their site, as well as wasted time and money during the project. It could also store up problems for the future if defective work is carried out, and there could be significant costs to put the foundation work right once the houses have been constructed on top of them.

To avoid these problems, the principle may be applied to construction management using the following steps:

1 Divide up the work – break the work down into skills or tasks involving different plant and materials, in this case setting out, excavation, reinforcement, concrete.

2 Use skilled workers for the different tasks – although the work will be carried out by a team of groundworkers, they may have different specialist skills among them. Good construction managers know their workforce and know who would be best to work with the machine during the excavation and who would be best to check and manage the concrete – this will improve the quality of the work and eliminate defects.

3 Specialisation – allocate workers to specific tasks to be carried out on all the plots; this increases their efficiency and maintains consistent quality.

4 The construction manager can now sequence the work – rather than inefficiently completing one plot before starting the next, work can be planned so that tasks will be carried out sequentially on the plots. For example, once the setting out on Plot 1 is complete, the excavator can move in, once the excavation is complete the reinforcement can be fitted, and once that is in place levels can be established and the concrete poured. Work tasks can follow each other round the site, keeping everyone working at once, but in sequence and safely.

Breaking down the tasks of 20 foundations along this principle of management can prevent problems from arising as well as help to ensure that quality work is carried out. Management of labour, plant and materials becomes simpler, as there are clearly allocated tasks which can be planned, coordinated and controlled. Applying the division of work means that the whole task of 20 foundations can be completed as effectively and efficiently and safely as possible.

Exercise A

Consider the remaining principles as they are detailed in Table 2.3.2. For each principle, consider the scenario and think about the following:

- What could go wrong if work was not carried out following the principle?

- How could construction managers implement the principle?

Suggested solutions to Exercise A may be found in the Appendix.

While the principles of management may seem a little disjointed from what construction managers actually *do*, from the above exercise it may be seen that they can still be very helpful in explaining *how* they should do it. The principles are relevant to many aspects of construction management, and provide a theoretical framework to help understand how people respond to and carry out activities as part of their everyday work.

2.4 Managing people: the key to successful management?

The main asset of any construction company is its people. Construction work is highly social – people come together, bringing their different knowledge and skills, to work as a **team** to achieve project success. Managing people is therefore an essential aspect of construction management. Construction managers must understand the social aspects of their work, and try to manage people accordingly.

Table 2.3.2 The principles of management explained

Authority and responsibility – Managers must be able to give orders, which means that they are also responsible for them.
Discipline – Employees must follow the rules, but there should be careful use of punishment.
Unity of command – Each employee should only receive orders from one superior.
Unity of direction – One manager in control of all the activities leading to one final objective.
Subordination of individual interests to the general interest – Everyone should be working together towards the common final objective, not their own individual goals.
Remuneration – A fair day's pay for a fair day's work.
Centralisation – Decisions about work can be kept centralised with the manager, or decentralised to others; a balance is needed for the right decisions to be made by the right people.
Scalar chain – The management chain. The level of authority reduces down the scalar chain, and there must be clear communication in both directions. It can be broken if this is in the interests of the business.
Order – A place for everything, and everything in its place, including labour, plant and materials.
Equity – Managers must be fair and just to their employees.
Stability of personnel – Low rates of labour turnover mean more efficient production.
Initiative – Employees should be able to bring new ideas to their work.
Esprit de corps – Team spirit!

There are many theories that explore why people behave as they do in different situations, and there have been many books produced around these different theories. In this section we will look at four of the most important elements of people management in detail:

- **Communication** is essential to ensure that information is passed on and understood.
- **Leadership** is vital to ensure that everyone is working towards the same goals.
- **Motivation** is needed to inspire commitment from everyone on the project team and to inspire them to produce their best work.
- **Teamwork** is what brings this all together.

A key part of the construction manager's role is to create an effective and efficient project team from all the different professions and trades involved in the project, ensuring that they are communicating effectively and want to work to the best of their ability as individuals as well as together, and to share their skills and knowledge to achieve project success.

2.4.1 Communication

Within modern management thinking communication is seen as essential. It is often added as number seven to Fayol's six management functions. Communication was identified as the most important management function for construction managers by Griffith and Watson in their survey (2004: 33).

Communication supports all other management functions and principles. The information and decisions made at each stage must be communicated effectively for them to be implemented correctly. Effective communication is not just about passing information on, it is about making sure that it has been clearly *received* and *understood*.

For example, after explaining the day's work to a team of groundworkers, the construction manager should not just ask if the instructions have been understood – the answer will usually be 'yes', just so that they can get out of the office and on with some work!

To make sure that the plan has been clearly *communicated*, a good construction manager will ask the team to discuss it and explain details of how they plan to organise the tasks to make sure that they are using their time most effectively and efficiently during the day.

Discussion Point

In what ways can people communicate on site?
How could problems of misunderstanding arise?

Poor communication is often identified as a key problem in the industry. Projects involve lots of different people, and long supply chains which need to share considerable amounts of information for project success. This requires well-organised networks of communication that use the most appropriate (if not necessarily the latest) technology.

Often communication is overcomplicated; too much information is passed on, resulting in floods of paperwork or emails, or too little, and time is wasted looking for the information that is really necessary. On site it is vital that information is up to date. For example, drawings can quickly go out of date and become superseded as new drawings are produced. Relevant information must be transmitted both verbally and in written form, and be clear, concise and fully understood to ensure that the right action is taken in the right place and at the right time.

The construction manager must be able to send this out efficiently and clearly to the relevant people on the site. This means that they must be excellent communicators in both directions, and be able to receive as well as pass information on effectively.

Exercise B: Scenario

You are working as the construction manager for a main contractor, and under your current design-and-build contract you are involved with managing the design team. While the architect and engineer seem happy to work together and share information, they haven't always included the mechanical and electrical (M&E) services designer. This has already led to one costly redesign of the plant room.

- What problems could have arisen here?
- In your role as the manager, how would you try to solve them?

Suggested solutions to Exercise B may be found in the Appendix.

2.4.2 Motivation

To achieve an effective and highly performing workforce, people must be motivated to work to the best of their ability. Therefore it is essential for construction managers to be able to motivate all members of their construction team.

There has been plenty of research investigating motivation; it is one of the most researched and written-about topics in management theory literature. This is not very surprising, as the answer is really the key to company success. Consequently a number of theories and ideas have developed to help managers understand how to motivate the workforce.

Discussion Point

What do you think motivation is?
Can you name some examples of what motivates people at work?

Put simply, motivation is the reason or reasons behind someone's actions or behaviour – this can be either a carrot or a stick. For example, a bonus can motivate people to work faster – a carrot – whereas reduced or 'docked' pay for slow work – a stick – could also encourage them to hurry up. Everyone is motivated by different things. Although money is the most common motivator used at work, people are also motivated by teamwork, being part of a recognised trade, or simply the satisfaction in producing good work.

Discussion Point

What motivates you?
Can you come up with a list of top five motivators?

Construction managers need to find a link between an individual's personal motivators and the effective performance needed to meet the project objectives.

For example, a joiner's personal goal may be monetary reward or a bonus, while the goal of the organisation is the completion of the project on time, on budget and to the correct levels of quality, health, safety and environmental control. Therefore the construction manager must set the bonus to an amount which optimises the joiner's output, but also makes sure that they meet the project requirements, encouraging their performance so as to be as effective as possible.

Notice that the above scenario is about optimisation of output within the parameter of the project requirements; it is not just about speed. Faster work can mean a reduction in quality, which would not be *effective* performance if work had to be redone in the future. The construction manager needs to make the joiner's personal goals line up with all the goals of the project.

Discussion Point

How could you link this bonus to the wider project goals? What would you include in the bonus calculation?

Theories of motivation

Theories of motivation help us to think about motivation in different ways, and to remind us that people are all very different and may therefore be motivated by very different things. No one approach will have the same motivational effect on everyone on site and construction managers need to be aware of this, and to have different approaches available – both sticks *and* carrots.

Maslow's needs are often shown as a triangle, as illustrated in Figure 2.4.1.

Maslow's hierarchy of needs

This is one of the most famous theories of motivation. In the 1950s, Maslow suggested that human needs operate at a variety of different levels, from basic physiological needs, such as hunger, to higher level needs, such as self-development and self-fulfilment.

Maslow's general theory was that, all other things being equal, people tend to satisfy their lowest level of need and then move on upward to the higher levels. If there is a problem with these lower levels then people focus on them and they are not able to move up to the higher levels.

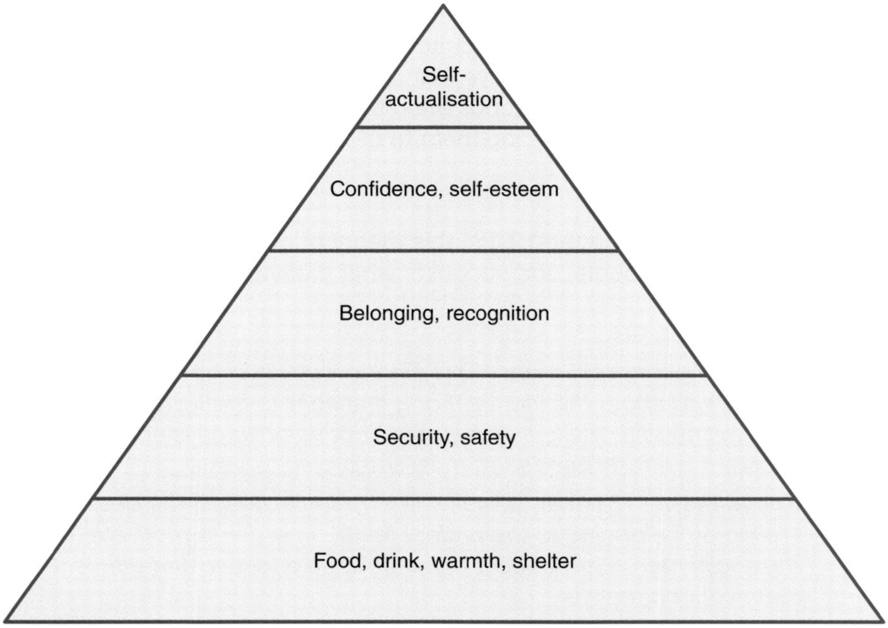
Self-
actualisation
Confidence, self-esteem
Belonging, recognition
Security, safety
Food, drink, warmth, shelter

Figure 2.4.1 Maslow's Hierarchy of Needs

Discussion Point 2.10

Can you relate each of these levels to the role of the construction manager?
For example, what would happen if the power to the site canteen went off?

McGregor's Theory X/Theory Y

This theory was put forward by McGregor in the 1960s, when he suggested that traditional management, the use of money and discipline (the carrot and the stick) as motivators, was not the only or even the best way to achieve results.

McGregor put forward a theory, based on two types of people, called Theory X/Theory Y:

1. **Theory X** argues that people do not like work; they must be forced to get on with it, and left to their own devices will not make the effort. They do not want to be involved or take on any responsibility for work or its management.
2. **Theory Y** argues that people do like work; they like to participate and be involved. They are able to self-motivate and enjoy the efforts of thinking, problem-solving, producing outputs, achieving goals and simply taking part.

Implementing Theory Y as a manager means spending time engaging and including people within work activities, and may be seen as a softer option than a more authoritative managerial approach. Yet research has found that Theory Y approaches often motivate the workforce far more than giving orders and forcing them to be carried out.

Schein's theory

Devised by Schein in the 1970s, this theory of motivation identifies four main "types" of people, and suggests what motivates them:

Rational economic man – This man is motivated by personal gain; it's all about the money. For a manager to motivate this man they will need to make sure that their bonus or wage is able to secure the efforts needed to meet the project goals.

Social man – This man is motivated by social needs, the desired feeling of 'belonging to the project team'. In order to motivate the social man, the manager must facilitate effective and efficient teamwork on the site.

Self-actualising man – This man is motivated by self-fulfilment, solving problems and taking on challenging work. Managers need to motivate this man by giving them responsibility and making sure they can take a sense of pride in their work.

Complex man – This man is complicated! Motivation for this man will vary depending on the time and place, and so the manager must be able to put the right motivator in place at the right time.

While Schein's theory supports the fact that money is a key motivator for many people, there is also the possibility that they are motivated by other factors and in some cases by different factors on different occasions. This means that construction managers have to be able to establish, through good communication, just what type of 'man' they are dealing with, and then use that knowledge to implement the most effective motivators to ensure that personal and project goals align.

2.11

Discussion Point

Can you identify Schein's four men within McGregor's theory? Are they X or Y, or a mixture of both?

De-motivation

As we have seen in these various theories of motivation, there are many different suggestions as to what will motivate people. For construction managers this variety can itself be a problem: how can you motivate all of the people all of the time?

The answer to this is that it's probably not possible.
But what can be done is to concentrate on removing any **de-motivators**. Developed by Hertzberg, and known as Hertzberg's Motivation-Hygiene Theory, this suggests that while one set of job characteristics can motivate people, another set can de-motivate them. These characteristics are closely linked to Maslow's hierarchy of needs.

For example, if people are dissatisfied with their working conditions they are not likely to be motivated to work to the best of their ability, as is shown in Figure 2.4.2.

If a groundwork gang are not able to dry their work clothes overnight in an effective drying room, they are not likely to want to start again bright and early the following day. It is up to construction managers to ensure that the welfare on site is to a good standard so that it does not become a de-motivator for the workforce.

What is a motivator can often become a de-motivator if not managed correctly; for example, any self-actualising man who takes pride in their work will become de-motivated if they are told to use poor-quality materials, or they are kept waiting for materials to arrive. Poor management and a lack of communication can also become de-motivators.

Discussion Point

What do you think are the most common de-motivators on sites? How would you try to resolve them if you were the construction manager?

Being able to motivate the workforce effectively is a key part of construction management. The theories outlined here may be applied to practice, to explore the best approaches to motivate individuals, but with the recognition that money may not always be the best solution.

Exercise C: Scenario

You are a construction manager and have been tasked with managing the joinery package on site. However, one of your joinery gangs is not meeting the required outputs in fixing the doors on the first floor, and you can see that they are falling behind programme. When you speak to the gang they tell you they are making enough money with the bonus rate they are being paid, and there isn't any need to go faster. However, if they don't meet the programme this will have a knock-on effect on the following decoration trades.

- What motivational factors are in play in the above scenario?
- How could you resolve the issue to prevent it from affecting the programme?

Suggested solutions to Exercise C may be found in the Appendix.

Figure 2.4.2 Poor working conditions can be a demotivator

2.4.3 Leadership

As with management, there has also been a large amount of research around leadership, and as a result there has yet to be a fully accepted definition of what leadership actually is.

Leadership is not the same thing as management – where management ensures that the work is carried out to meet the company's objectives, leadership is involved with having the vision to inspire others to want to work towards those objectives. Leadership is closely linked to motivation, and leaders are types of people who are able to motivate others.

Fryer *et al.* (2004) suggest that leadership is hard to define because it is so complicated, and involves many intangible things that cannot always be defined and measured. For example, leadership is about setting goals for the future and inspiring others to work towards them – something that is hard to define in terms of how it can be done effectively. Leaders often emerge when they are needed to help a group achieve a specific goal – and it should be the aim of the construction manager to be the leader for their project, to inspire people towards the goal of project success.

Discussion Point

Who is the leader on a construction site?
Is it always the construction manager?

There are frequently both formal and informal leaders within a company or on a site.

The **formal leader** would be the person who has been appointed to the role – the construction manager, for example. As well as fulfilling their management functions, companies often expect managers to *also* perform the role of leader, and to achieve high productivity and satisfaction on their projects.

Alternatively, **informal leaders** may emerge. This is the person to whom the group will turn when problems arise. It can also be the person with the most experience of a situation or particular work practice, and who is able to provide insight as to how to complete or solve the problem.

One theory is that leaders are born and not made. Effective leaders are born with innate qualities and are destined to lead, so when looking to appoint people as managers, certain traits and skills should be identified.

There are considered to be four main identifiable leadership traits:

- Staying calm under pressure
- Owning up to mistakes
- Good interpersonal skills
- A wide range of knowledge.

Discussion Point

How valuable are these traits in leaders within the construction industry?
Can you think of any more desirable traits?

This way of thinking suggests that leadership cannot be taught and instead is something that already exists within certain people. Another school of thought believes that through experience people can develop the skills to become leaders over time.

Some people naturally become leaders within any given situation. It is highly beneficial for construction managers to be good leaders, to support their effective management of the project towards successful completion. Whether managers should be appointed because of their leadership skills and learn their management skills on the job, or whether they should be appointed because of their management skills and develop leadership skills over time, is an ongoing debate.

Exercise D: Scenario

You have just been transferred to a new domestic development site to take over from another construction manager who has left the company. While familiarising yourself with the site, you meet the brickwork foreman who seems to be in charge of most of the works on site. While you are talking to him, several other trades come up to ask for information or advice, and he seems happy to help. However, you notice that some of his ideas may help to complete the brickwork quickly and efficiently, rather than the houses and the project as a whole.

- How could this shift in leadership from the formal management team have occurred?
- What problems could this cause you in the future as the construction manager?

Suggested solutions to Exercise D may be found in the Appendix.

2.4.4 Teamwork

Construction by its very nature is a team-based industry. Modern construction projects cannot be delivered by individuals; people have to work together in multidisciplinary teams to achieve success. Therefore understanding teamwork and how effective teams work is vital for construction managers.

There are a variety of teams within construction projects. For example, the client's professional team consists of architects and other designers, engineers, surveyors and project quantity surveyors who must work with the site team, including managers, engineers and the subcontractors who actually carry out the work. Both of these teams need to work together to achieve success, an understanding which has seen the development of **partnering** and **collaborative working** processes, a shift away from old-fashioned, adversarial approaches to construction.

The advent of **building information modelling** (BIM) brings the potential for further collaboration and improved teamworking, as project teams become more integrated through the BIM processes and the development of shared design models. BIM will mean that people *have* to work together through the shared model and shared work practices, which will hopefully benefit the whole project. More about BIM and how it is changing the way we work may be found in Part 4.

Individuals

Teams are made up of individuals. Managers therefore need to be able to understand how and why individuals behave and perform at work, so that they can try to predict how they will perform as part of a team. This links closely to individuals' motivation and how they respond to leadership.

One idea suggests that people behave the way they do because of an inherent 'personality'. Personality may be considered to be the set of characteristics which determines the sort of person they are. We tend to label people as types – friendly, hostile, shy or domineering – if they usually display that kind of behaviour.

Discussion Point

What characteristics do you think are necessary for good construction managers?
Why?

In fact people can behave very differently in different situations; personality is not rigid, rather it is constantly changing. People are influenced and affected by their interactions with others and the situations in which they find themselves. Such influences and interactions are an inherent part of being part of a team.

The processes and dynamics of the team can be highly influential on individuals, and managers must appreciate them in order to be able to manage them effectively.

Teams and groups

There are a number of management theories that have developed around groups of people which are very relevant to the construction industry due to the large number and variety of groups that are created or which form naturally within the work environment.

For example, groups may be defined as either formal or informal:

Formal groups are set up by management to undertake specific tasks – such groups are called **teams**. One example is the project team, which is established to achieve the specific goal of project success.

Informal groups develop to fulfil a need within the workforce; for example, employees who enjoy sport may form an informal five-a-side football team outside of work.

2.16

Discussion Point

What formal groups can you identify on a construction project?
What informal groups might also be present?

One theory suggests that all groups tend to have the same main attributes:

- **Norms of behaviour** – members of the group will conform to a certain type of behaviour.

- **Identity** – the group gives its members a specific identity which will be understood by people outside the group.

- **Hierarchy** – the group has some form of 'pecking order'; this may be agreed by the group or imposed upon it.

- **Exclusivity** – the group has the power to allow new members to join and the power to remove existing members.

- **Solidarity** – members of the group stick together. Although there may be arguments or disagreements within the group, it will always form a united front externally.

These group attributes can either help or hinder management and influence how the group or team works with other groups or teams. For example, group solidarity can be a problem when different **gangs** have to work together on site – one bricklaying gang may be very keen on quality (a norm of behaviour) while another gang may be less concerned. This can cause problems for the construction manager who has to maintain high-quality standards throughout the project.

Group norms of behaviour are very important, and understanding their influences is essential for construction managers. Many people will conform to group norms, even when they conflict with individual preferences. Some people will deviate from group norms, but many people simply behave as they think others expect them to behave.

Discussion Point

2.17

How could the influence of group norms affect how people behave on sites?
Think about safety, quality, working hours, attendance and work ethic.

Teams are simply groups with fixed goals, and so teams, such as the site team, may be influenced – both positively and negatively – by group norms of behaviour and the development of informal groups within the same social setting.

Construction managers must ensure that teams on site, from the operative gangs to the project team as a whole, have positive attributes and norms of behaviour to help achieve project success.

Exercise E: Scenario

You are the construction manager on a new-build domestic site. You notice that the new junior quantity surveyor is not pulling their weight and seems to spend most of the day outside the site in the smoking area talking to the operatives. When you tackle the senior quantity surveyor they are reluctant to talk about it and don't seem concerned.

- What aspects of group theory could be happening here?
- How could you bring the junior QS back into the project team?

Suggested solutions to Exercise E may be found in the Appendix.

2.4.5 Managing people in construction: summary

The construction industry is all about people working together. Construction managers need to understand people in order to be able to manage and lead successfully. They need to know how people can be motivated or de-motivated at work, and how they can be influenced when they come together in the many different teams found in the construction industry.

The theories behind managing people may be used to help construction managers when investigating problems or developing new strategies and approaches to improve their own project performance, and to ensure that their own management style is efficient and effective when dealing with people.

2.5 What is construction management?

The role of the construction manager within the construction industry has often proved hard to define, possibly because it has to change as the industry changes in terms of management processes and technologies.

Fryer *et al.* (2004: 1), in their book *The Practice of Construction Management*, suggest that there are actually two *types* of definition of construction management.

The first is straightforward:

Construction management is about getting things done properly throughout the whole construction project.

This sums up the ultimate goal of the construction manager.

The second type is more complicated and includes the different processes that need to be carried out by construction managers, such as forecasting, planning and controlling work, as well as *how* these processes should be carried out. Within the modern construction industry Fryer *et al.* (2004: 2) argue that the amount of communication, regulation, technical knowledge and cooperation necessary for managerial success means that construction management has actually become a team game for large companies – it's too complicated for only one manager to deal with all aspects of modern construction operations.

This definition aligns with the theories explored within this section, including management functions and the principles of management, as well as the need for people management and teamwork.

Griffith and Watson (2004: 31), in their book *Construction Management Principles and Practice*, specifically define construction site management, but again there are many different aspects to consider. As well as conducting and controlling the work, they state that construction site managers also need to be able to:

- lead their team;
- have a sound technical knowledge of construction;
- be dedicated to their work;
- be experienced;
- have a strong character;
- work well under pressure.

These characteristics are again grounded in the theories around managing people, and also refer to more personal traits of the individual managers themselves.

The Chartered Institute of Building has also produced an *Inclusive Definition of Construction Management*. As its title suggests, all levels of construction management are incorporated, from the **site** to **project** to **corporate** management, and all phases of the construction process from 'cradle to grave'. Therefore, this definition is **strategic** in its perspective, namely an overarching idea of construction management rather than a detailed prescription of tasks:

> Management of the development, conservation and improvement of the built environment; exercised at a variety of levels from the site and project through the corporate organisations of the industry and its clients to society as a whole; embracing the entire construction value stream from inception to recycling, and focusing on a commitment to sustainable construction; incorporating a wide range of specialist services; guided by a system of values demonstrating responsibility to humanity and the future of our planet; and informed, supported and challenged by an independent academic discipline.
>
> (CIOB 2010)

Although this definition does not draw explicitly on any management theories, it clearly shows the complexities of the industry by placing the role of the construction manager firmly in context.

2.18

Within the construction industry, there is not just one definition of construction management that covers all perspectives. Various ideas have been put forward, which start from the very simple and become increasingly complicated as more detail is added. In the modern construction industry the role of the construction manager is actually very complicated.

Discussion Point

How would you define construction management? Try to fit it into just one sentence.

But despite the lack of only one definition, and the need for more and more detail as people try to explain how to do it, there are some shared ideas within these different definitions that may be used to help us understand construction management.

Construction is a very complex industry; many people come together on a project-by-project basis to construct bespoke outputs to specified standards. This process draws on the functions of management to set out what managers actually do, the principles of management to implement these functions within project teams, and theories of people management to ensure strong leadership, workforce motivation and teamwork to ensure project success.

On a practical and simple level, construction management may be considered as making sure that **people** are delivering what they should be delivering, on **time**, to the required **quality**, within the set **budget**, to the necessary **health**, **safety** and **environmental** standards.

Managers must be **efficient** and **effective** to meet these requirements, but also be **flexible** enough to change with different situations, different types of project, different project teams, different procurement routes and different ways of working.

2.6 Construction management in theory: summary

The theory of construction management is firmly rooted in wider theories of management. What managers do is grounded in Fayol's management functions of forecasting, planning, organising, coordinating, motivating and controlling, which define the key activities managers should undertake in their work.

How managers should undertake their activities is explained through Fayol's management principles, which provide a theoretical basis for the application of management to people at work. Managing people is one of the most important aspects of management; people are vital to the construction industry, so leadership, motivation and teamwork are vital too. Construction managers must be able to understand how people may react to different events in different situations in order to ensure that they are getting the best out of their people, while keeping them happy and healthy at the same time.

Part 3 of this book aims to support the development of your theoretical knowledge and understanding of management, and to explore exactly *how* construction managers carry out their work on a daily basis. Construction management in **practice** is introduced and explored; and the ways production is planned and controlled, and how sites, time, cost, quality, health and safety and the environment are managed by construction managers on a daily basis.

Further reading

Management theory is explored in more detail in the following texts:

Fryer, B., Egbu, C., Ellis, R. and Gorse, C. (2004) *The Practice of Construction Management*, Blackwell, Oxford.

Griffith, A. and Watson, P. (2004) *Construction Management – Principles and Practice*, Palgrave Macmillan, Basingstoke.

These textbooks provide the next steps in developing your understanding and knowledge of construction management theory and practice, and both include chapters on leadership, motivation and teamwork.

You can also find out more about many different aspects of construction management in *Construction Manager* magazine, published by the Chartered Institute of Building.

Discussion point comments

Discussion Point

What do you think management is?
Can you explain it in just a few words?

This is a hard task!

Most simply, management may be seen as the coordination and control of people at work.

Discussion Point

What might directors of large construction companies be trying to forecast?
What might construction managers be trying to forecast?

A company director may be looking at past industry trends to inform the next 10 years. This could follow government investment plans; a shift to infrastructure investment could mean that the company should look to develop this area of their work. Alternatively, the director may be looking to the wider global economy for trends of growth or austerity, and whether the company should look for projects overseas if there is not enough work in its home country.

A construction manager may be looking at the overall duration of their project and the workload over this period of time, and whether there may be issues in relation to the key resources of labour, plant and materials. For example, a sharp increase in house building can result in a lack of both bricks and bricklayers within the marketplace, and so costs for this element of their project could rise.

Discussion Point

Why does construction work need to be planned?
What elements would the overall project plan need to take into account?

Construction work is very complex. Large numbers of people, **subcontractors**, plant and materials must come together at the right time and in the right order and in the right way to meet project requirements. The use of subcontractors and subcontracting within the industry also means that many different plans need to be coordinated to meet the ultimate goal of **project success.**

A good plan at any level must be flexible, based on the most up-to-date and accurate information, and easy to understand. Milestones or goals need to be included to make the plan relevant, such as the start and end dates of the project, and **resources** must be accurately and efficiently employed to meet them. Construction management resources are usually the labour, plant and materials needed to complete specific tasks.

2.4

Discussion Point

How does the use of subcontractors influence the allocation of tasks at site level?

What effect does this have on the coordination role of the construction manager?

The use of subcontractors splits the work into specialist areas, which are assigned as **work packages** and which specify the scope and details of the work. This means that work is allocated to different subcontractors before the works commence on site, but these work packages often have interfaces with each other and so the different subcontractors will rely on each other too – for example, the windows cannot be fitted until the walls are built!

For the construction manager, this means they have to coordinate a large number of different subcontractors in both time and physical space. But all of these subcontractors want to get their work package completed as effectively and efficiently as possible – which does not always meet the overall project objectives. The construction manager must ensure that the subcontractors work together as a team to meet the overall project requirements.

Further complications may arise when the subcontractors also subcontract elements of their work down the supply chain, and so even more different subcontractors are working on the site, again with their own separate goals. It is also assumed that the work packages which have been assigned do not contain any gaps in the work, such as pipe boxing or mastic, which are needed to complete the project.

2.5

Discussion Point

What corrective action could be taken if performance is not meeting the original plan?

What corrective action could be taken if cost is not meeting the original plan?

If performance is not meeting the plan, the construction manager needs to increase productivity. This may be done in a number of ways, including the following:

- increase the workforce;
- improve the bonus system;
- revise the plan;
- change the working hours.

If cost is not meeting the plan, the construction manager needs to reduce spend. This may be done in a number of ways, including the following:

- revise material specifications;
- change materials supplier;
- value engineering (amend the design so it is cheaper to produce);
- change sequence of work or supervision to accelerate production.

Discussion Point

2.6

In what ways can people communicate on site?
How could problems of misunderstanding arise?

People communicate on site in many different ways, including the following:

- face-to-face communication, including verbal and non-verbal gestures;
- written communication through posters, noticeboards and memos;
- mobile phones and texts;
- the programme of work;
- through drawings and sketches.

Misunderstandings may arise if the communication is not clear and/or is not understood by the recipient. This could be due to lack of clarity in drawings or **specifications**. Language can also cause problems on site; operatives from different countries may not speak English fluently, but also many of the terms and words used in construction can be confusing even to people for whom English is their first language – for example, do you know where the 'rubber duck'* is?

Discussion Point

2.7

What do you think motivation is?
Can you name some examples of what motivates people at work?

*a rubber duck is a dumper with inflatable tyres.

Motivation is the reason why people do things.

People can be motivated by money, recognition, the feeling of a job well done, being part of a team, seeing your work make a difference, the possibility of promotion, long holidays, short working hours, flexible working patterns, job security or reasonable travel to work.

2.8

Discussion Point

What motivates you?
Can you come up with a list of top five motivators?

Many people think they are only motivated by money!
Your top five motivators may include those listed in the answer to discussion point 2.7, as well as others.

2.9

Discussion Point

How could you link this bonus to the wider project goals?
What would you include in the bonus calculation?

Bonus is often just paid for speed of production – the more work carried out the larger the payment. However, it is possible to add in other elements to support the project goals. For example, to ensure that increased speed does not reduce quality, construction managers can factor in a portion of the bonus for quality of work. Bonuses may also be structured to incorporate acknowledgement of health and safety and environmental performance.

2.10

Discussion Point

Can you relate each of these levels to the role of the construction manager?
For example, what would happen if the power to the site canteen went off?

At the lowest level of Maslow's pyramid, the construction manager needs to make sure that the workforce has shelter, a dry, warm and comfortable place to eat, good, clean toilets and washing areas – all of these aspects form part of the welfare provision on site. The construction manager should also make sure that people have been provided with the necessary wet weather clothes to keep them comfortable at work.

The next level requires safety and security, which can mean both physical safety on site through effective H&S management and provision of personal protective equipment (PPE), as well as security in terms of job security and future plans.

The next level is to do with teamwork, and the construction manager must ensure that the workforce feel part of the construction team and are given appropriate recognition for work well done.

Beyond that the manager can support people in developing confidence, and help them to achieve self-actualisation and satisfaction through their work.

The example given of a lack of power in the site canteen would mean that the workforce would not have a warm, dry and comfortable place to eat their lunch, and if it was raining and cold this would negatively affect motivation and morale. Maslow's theory states that if a lower level of the hierarchy is not satisfied people cannot move to the next level. Here, the lowest level has not been met, which would mean that people are not receptive to the higher levels to motivate them – it is likely that very little or poor-quality work would be done that day.

Discussion Point

Can you identify Schein's four men within McGregor's theory? Are they X or Y, or a mixture of both?

Rational economic man fits with McGregor's Theory X – they are just motivated by money and do not want to work for any other reasons.

Social man and self-actualising man fit with McGregor's Theory Y – they are motivated by other interests such as their own commitment to the tasks of work.

Complex man is a mixture of both, and may show characteristics of both Theory X and Theory Y.

Discussion Point

What do you think are the most common de-motivators on sites? How would you try to resolve them if you were the construction manager?

The most common de-motivators will often depend on the situation; for example, in winter cold cabins with nowhere to dry clothing will be a big de-motivator.

De-motivators may include the following:

- unachievable bonus;
- low basic wage;
- poor working conditions;
- lack of materials;
- lack of sufficient welfare;
- unsafe working environment;

- poor-quality PPE that is ineffective;
- re-work owing to poor planning or organisation;
- re-work owing to poor communication;
- poor site layout (distance from stores to workface);
- lack of security;
- work damaged by others – poor trade sequencing.

Many of these de-motivators should not arise if construction managers plan the sites and the work correctly. For example, welfare provision should be made to suit the maximum number of workers on a project at any given time, PPE should be fit for use and not break after one day, and work should be planned to ensure that it is able to be completed 'right first time' and protected from damage by following trades.

Discussion Point

Who is the leader on a construction site?
Is it always the construction manager?

The leader on the construction site may be the company owner, the construction manager, or even the general operative in charge of keeping the site in order. Leaders will be the individuals to whom people turn when they need guidance, and so the general operative may well be the leader on a practical level, as they are the one keeping people motivated and moving forward.

On site, the leader should be the construction manager, as they are in charge of the site and its operations and so should have a good presence and be recognised by the workforce as the leader, able to quickly solve problems and motivate people to work to the best of their ability. But this depends on their approach. Some construction managers just lead their site management team, some lead the whole site team, and some rarely come out of the office and so don't really inspire leadership on site at all.

Discussion Point

How valuable are these traits in leaders within the construction industry?
Can you think of any more desirable traits?

These traits may all be considered valuable for leaders within the construction industry; for example:

- Staying calm under pressure – construction sites can be hectic places, and leaders need to be able to stay calm under such circumstances.

- Owning up to mistakes – blaming others for poor decisions or mistakes will alienate leaders from the site team, and mean that they may not be able to count on people in the future, as people may feel vulnerable to blame if something goes wrong again.
- Good interpersonal skills – construction is a people industry, and communication and the ability to persuade, motivate and encourage people to work to the best of their ability to achieve the project goals is vital for leaders.
- Wide range of knowledge – leaders should be able to see all sides of a problem, and understand the different trades and technologies in order to be able to discuss the issues in detail and come up with the correct solution.

Some more desirable traits could be the following:

- decisive;
- able to manage stress;
- fair;
- confidence to delegate;
- consistent;
- able to motivate.

Discussion Point

What characteristics do you think are necessary for good construction managers? Why?

Some characteristics could be as follows:

- Decisive – able to make a decision and stick to it so that people can get on with their work.
- Able to manage stress – and to deal with the pressures of the construction site.
- Fair – to treat everyone equally.
- Experienced – to draw upon experience to find solutions to problems and foresee the outcomes of their decisions.
- Confidence to delegate – construction is a team game and the manager must be able to share the workload.
- Consistent – changing minds and decisions can be costly on site and result in re-work, which is a de-motivator for those having to carry it out.
- Able to motivate – to inspire the workforce towards the goal of project success.

Discussion Point

What formal groups can you identify on a construction project? What informal groups may also be present?

Formal groups may include the following:

- professionals;
- the project team;
- the client's team;
- site supervisors;
- foremen;
- different trades/subcontractors.

Informal groups may include the following:

- those who eat in the canteen rather than go off site;
- those who work in certain areas;
- site sports teams.

Discussion Point

How could the influence of group norms affect how people behave on site?

Think about safety, quality, working hours, attendance and work ethic.

Group norms influence how people behave.

If a group has a high regard for safety, then the group will wear their personal protective equipment (PPE) and follow the safety rules. If there is little regard for safety, people may be more likely to take risks in their work and not bother to wear their PPE.

Some groups pride themselves on the quality of their work, and so will strive for the highest quality output. Alternatively, some groups are keen to work as fast as possible, and quality is seen as something to make good later on, producing large quantities of poor-quality work.

If a group arrives early and leaves late on site, then new members will also follow this pattern. Alternatively, if people come in late, leave early and only work half a day on Fridays, then this can also influence new members and cause problems for the construction manager, as the work outputs will have been planned for a longer working day.

Some groups will think it acceptable if members do not attend regularly; for example, if they are working on other projects at the same time, or simply the morning after a football match the night before. If sporadic attendance becomes acceptable this can again cause problems in terms of planned work output.

All of the above are closely linked to the idea of work ethic, and these factors therefore influence the group norms. Construction managers need to inspire a strong work ethic on their projects and encourage this to filter into the group norms of the different formal and informal groups on their sites, in order to ensure effective and efficient working.

2.18

Discussion Point

How would you define construction management?
Try to fit it into just one sentence.

This is another very hard task.

Construction management may be defined as making sure that people are delivering what they should be delivering, on time, to the required quality, within the set budget, to the necessary health, safety and environmental standards. Management must be efficient and effective in order to meet all of these requirements.

This is actually two sentences!

PART 3

Construction management in practice

Introduction

Part 3 is all about construction management in practice. This focuses on the management of **production**. It aims to develop an understanding of the different elements that contribute to and affect production in construction industry operations, and how they are managed on sites.

It will look at the following subject areas:

- What production means in the construction industry.
- The parameters that affect production management and how they are interrelated.
- The two sides of production management: planning and control.
- Site management in terms of **site layout** and daily operations.
- How time is managed on sites.
- How costs are managed on sites.
- What **quality** means in construction, and how it is managed on sites.
- Why **health** and **safety** is such an important part of management practice.
- How **environmental** impact is managed and mitigated on sites.

Construction management in practice covers a wide variety of different specialist areas of management, and various theories and approaches have been developed by both academics in their research and construction managers as they work out on sites. Key terms and ideas are introduced, as well as knowledge about how they are carried out, to give you a strong grounding in practical construction management.

For more information about the construction management skills introduced here, and to find out about more complex tools developed from them, several more advanced textbooks have been recommended in the further reading section to help you build on the knowledge gained here.

3.1 Production planning and control

Production is the 'bread and butter' of the construction industry, it's what we do: we produce things. We produce houses, schools, hospitals, roads, bridges, tunnels, skyscrapers; in fact you could argue that production of the built environment is the function of our industry as a whole.

But we need effective management of this production to prevent projects from running over time, going over budget and resulting in poor-quality buildings. Effective production management also enables construction managers to ensure good health and safety and environmental management of projects. If these two elements are not considered within the overall production management planning and control systems, they cannot be managed effectively and will always be struggling to match the importance placed on time, cost and quality as the project is constructed.

Good production management is therefore critical to project success, and forms the vast majority of a construction manager's workload on a daily basis.

3.1.1 Production parameters

Production management involves managing all the elements that set parameters around construction outputs and can have either positive or negative effects on the production process and the end product. These include the following:

- The **time** it will take to design and build the project.
- The **cost** of the project, both the final amount and the cost for each month of production.
- The **quality** of the project, in meeting the specified design.

These three parameters are often seen as related to each other, and are often considered in the form of a 'Production Triangle', also known as the 'Iron Triangle', as is shown in Figure 3.1.1.

One way of looking at this relationship is to consider whether you would want a project to be excellent (high quality), fast (short **duration** or project time-scale) or cheap (low cost).

Old management thinking suggested that you could only optimise two out of the three, and one would always lose out. For example,

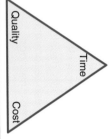

Figure 3.1.1 The production management triangle

if you want your project to be fast and excellent, then it will cost a lot to build; if you want it fast and cheap, then the quality will likely be poor; and if you want excellent quality and cheap, it will take a long time to design and build. Changing the objectives of the project influences the parameters accordingly.

This idea is useful in showing how these parameters are interrelated. However, in modern construction management other aspects should also be considered which have an influence, making production management even more complicated.

We should also include the following:

- The **environmental** management of the project, how 'green' the project is in terms of meeting environmental targets, such as low carbon production.
- The **health** and **safety** management of the project, with reference to both the workforce and the general public.

These are two key aspects of construction management, and they are examined in more detail later in this section. These parameters also contribute to and influence the construction outputs, and therefore production. Therefore they must also be included in our 'model' of production management, as is shown in Figure 3.1.2.

When we add these two parameters to the production triangle of time/cost/quality, we add more complications and considerations. For example, optimising the health and safety management on a project may mean that it will take longer to build if a safer but more time-consuming method of construction is chosen, or an alternative method may be as quick and safe but may cost more. This is the same for environmental elements; for example, if a project is designed to include more locally sourced and therefore **sustainable** materials, these may cost more than traditional, but less sustainable, construction products.

Discussion Point

What other relationships can you see between these five elements?
What are the different results of optimising the various elements?

This is not the only way of thinking about production management. New ideas suggest that *all* parameters can be equally optimised

Figure 3.1.2 The production management pentagon

through new approaches to construction management. For example, the approach of **lean construction** suggests that through good **supply chain management**, **collaborative working** and elimination of **waste** from the production process, projects can be delivered that meet all the necessary outputs without the need to compromise. Lean construction is discussed in more detail in Part 4.

But whatever management approach is used, the key parameters of production management remain the same. An understanding of how they exist in construction operations, how they interact with each other, and how they are planned and controlled remains a necessary part of good construction production management.

3.1.2 Where do these production parameters come from?

The production parameters are set by two sources: project clients and the government.

Time and cost are largely decided by clients, who usually specify both the speed and price they will pay for the project. They will dictate the broad time-scales for the project: the deadline for completion or whether several **sectional completions** and **partial handovers** are needed throughout the project. The latter can be common in residential projects where the client needs to sell some units to pay for the next phase of construction. The client will also dictate the total budget for the project through the tender price they accept. The total cost of the project may depend upon the **procurement route** chosen, but the client will need to be able to pay for the project on a monthly **valuation** basis as the works progress. Time and cost are also influenced to some extent by the practical nature of construction work. It takes a certain amount of time to complete work; mortar needs time to go off, concrete needs time to cure and buildings need time to dry out. Thus, although increasing spend often also increases speed, this is not always possible within the practical constraints of construction.

In the UK, quality is influenced by both the government and the construction client. All construction work in the UK must comply with the **Building Regulations** and meet certain standards, including British Standards, in both design and execution. These Regulations are easily available online and are managed and monitored by local council **Building Control** offices. The Building Regulations form the 'nuts and bolts' of the construction quality, ensuring structural stability as well as standards in terms of sound, heat transfer and fire safety. The client will also have input in the design of the project, which will be developed by the **architect** and **design team** to meet their individual requirements, as well as to comply with the Building Regulations. The design is then communicated through drawings and **specifications**, which detail all the materials to be used, how they are to be installed and what quality of finish is to be achieved. The client, through the design team, is able to influence the final look and finish of the building; what products and materials will be used for the cladding, windows and flooring. This is where the client input makes the difference, and the choices and level of finish and quality will be very different between, for example, a luxury hotel and an office block.

Health and safety and environmental management must also meet certain standards and requirements set down in UK legislation, the details of which are explored in more detail in Sections 3.6 and 3.7. However, the client may also set their own aims for the project, which may be linked to their company's wider goals. Some client companies pride themselves on their good health and safety record, others on how environmentally friendly they are. These goals are then applied to any construction work the client carries out and become production parameters affecting both the design and construction of their projects. For example, some clients may set their construction project an accident frequency target of zero, or the need to meet certain environmental targets, such as reduced water usage or waste recycling, in order to then support their own company in its corporate goals.

With all of these different parameters to consider and balance within the project delivery, production planning becomes a critical part of construction management in practice.

3.1.3 Production planning

As we saw in Part 2, planning forms Fayol's second management function. All managers will plan to different extents, even within the scope of only one construction project.

There are three key stages of construction management planning:

- Pre-tender planning
- Pre-contract planning
- Operational planning.

These reflect the various stages through which construction contractors progress when seeking to win and then carry out construction works.

When contractors are **bidding** for projects, and developing their **tender** for the work, they undertake **pre-tender planning**. Initially, this can simply be to check that the client's start and completion dates are practical and can be met for the project. They will also need to check that the project is achievable given their current workload and consider whether they have the resources to complete the project should they win, alongside other practical aspects such as the location and complexity of the work. Pre-tender planning involves the selection of optimum work methods and processes for the project, alongside consideration of health and safety and environmental issues, all of which contribute to an accurate and realistic tender price for the bid.

Once a project has been won, contractors then initiate **pre-contract planning**, developing their pre-tender plan into a more detailed proposal prior to the works commencing on site. Provision of a copy of this plan to the client before the work starts may be a contractual requirement, and the plan can become a key tool in the legal management of the project; for example, to demonstrate when work has fallen behind or been affected by considerations outside of the contractor's control, such as client changes to the design.

Once the contract has been agreed for the project, **operational planning** begins. This develops the pre-contract plan even further into a plan for the operational or construction phase of the project. This development is essential to ensure that the construction work keeps to the parameters of production that were specified by the client at tender stage and were included in the pre-contract programme.

The construction manager should be involved in all three stages of planning undertaken by the contractor. As the process of planning develops from the pre-tender stage through to the operational stage of the project, the construction manager's knowledge of construction at an operational level may be used to ensure that a realistic, practical and effective plan is produced from the very beginning, before the tender has even been submitted to the client.

For planning to be effective, it must be balanced to meet all the production parameters as specified for the project, but it must also be flexible enough to meet with changes should they occur.

3.1.4 Effective production planning

Gathering project information

Before effective production planning can be carried out, all the available information for the project must be collected.

The amount of information available at the pre-tender planning stage can be limited, but may include the following:

* Site location information (and ideally a site visit; see below).
* Ground investigation reports.
* Design information – drawings and specifications.
* Bills of Quantities (if applicable).
* The availability of resources, including key subcontractors who are often asked to contribute their own knowledge to the development of the pre-tender plan.

At the later stages of planning this information is supplemented by any documents produced during the planning process, and once the project has been won the contract documentation also becomes important in confirming the production parameters for the project.

The site visit

A site visit is essential. Construction work is carried out in real space, and this space may contain all kinds of things that could influence and affect any work. Although technology will now let us virtually walk around anywhere in the world, it may not show a new telegraph pole installed last week directly in the way of the site access.

A site visit should therefore check the following:

- the size and quality of the roads leading to the site;
- site access and turning into the access;
- potential security issues;
- who the neighbours are;
- existing structures on the site;
- the **topography** of the site;
- existing services (electric, water, drainage, etc.), both in the ground and overhead;
- what additional provisions may be needed.

Detailed records should be made, including photographs of key aspects of the site, and marked-up drawings to show where these features are within it. One tree can look very much like another if clear records are not kept while the photographs are being taken.

This information will contribute to the feasibility of the project in practice.

Essentially, it adds another project parameter in terms of the practical aspects of the project in real space, and this may also affect all the other parameters. For example, a weight restriction on the road to the site, as is shown in Figure 3.1.3, may mean that smaller wagons will need to be used for deliveries. This will mean smaller loads of materials, which may cost more in delivery charges.

Figure 3.1.3 A weight restriction on a road

Discussion Point

How could the findings of the site visit affect production planning?
How could they affect the different parameters of production?

The practical nature of construction work should not be overlooked. If a site visit is not carried out and a method of work is selected that cannot actually be used, then a change to the method at a late stage can be very costly for the contractor.

Selection of construction methods: how are we going to do it?

The planning process, through careful consideration and balancing of the production parameters alongside the practical information gained from the site visit, will result in the selection of the methods of construction for the project. This is initially made at the pre-tender planning stage and influences all future planning from that point onward.

For example, the method of construction for a multi-storey project could use either a tower crane or mobile cranes, concrete pumps and goods hoists.

The construction methods selected will determine the time it will take for the construction work to be completed. This **project duration** will be affected by the **resources** (labour, plant and materials) needed to complete the work using this method and in this time, which will in turn affect the cost of the project, which will also influence the ability of those methods to meet with quality requirements, as well as the health and safety and environmental goals of the project. Resource availability can therefore be a key consideration in itself; shortages of materials or skilled labour can affect the selection of construction method at a very fundamental level.

If the construction method should change at the operational stage of planning, then the resources priced for in the tender will no longer be valid, and the tender price may no longer be achievable. This may mean that the project will not be able to make a profit for the contractor. The new method may also take longer, increasing the overall project duration, or may not be able to meet the original quality requirements, which again will affect the cost and time of the project, as redesigns may be necessary. Pressure to make quick changes on site in terms of the practical work often means that health and safety and environmental concerns can also be neglected.

This is the kind of change that leads to problems of cost and time overruns for the project, and demonstrates the need for careful and considered planning at all stages. Construction contracts do allow for clients to make changes to the project if they are willing to pay for them, and allow extra time if needed. But if contractors need to make any changes through their own bad planning they cannot claim more time or money, and so work methods must be carefully planned from the very beginning to avoid problems arising.

Discussion Point

What could lead to bad planning by the contractor?
What would be the best way to prevent this from happening again?

The construction method selected will affect all the production parameters to some extent, and they in turn will influence the selection of the work method.

For example, a contractor building a speculative housing development will have to make a choice between using timber frame and traditional brick/block construction. Both of these methods of construction can be considered alongside the effect they may have on the production parameters.

Table 3.1.1 on page 76 shows that there are many considerations in planning the project around the production parameters. If time is highlighted, while timber frame is quicker to construct once on site it has a longer **lead time**, which may not suit the client's requirements to start the project very quickly; for example, if **planning permission** is soon to expire. Even if this is not the case, despite its short on-site construction period, the long lead time for timber frame may mean that the overall project durations would end up almost the same whichever method is chosen, so no time optimisation is achieved in reality.

However, practical considerations must also be taken into account, and this is where construction managers should use their knowledge and experience. For example, consideration must be made of the time of year; brick and block cannot be laid at 3°C or lower, and so work may be delayed if the project is planned for winter. Site access must be considered, as a tight or narrow access may mean that large timber frame panels cannot be delivered and smaller panels may need to be used. This could increase the costs and time for construction, again affecting other production parameters. Such practical considerations must be taken into account in order to ensure that the production planning will be effective once the project reaches the site.

In addition, you will notice several caveats in Table 3.1.1 – things must be *correctly specified and constructed* or *'managed correctly'* for them to influence the production parameters. This is where effective construction management is vital. All production parameters are vulnerable to poor management on site which can affect their planning and control, and may lead to delays, cost overruns, inferior quality construction, poor health and safety records, and environmental concerns.

Therefore all production planning is about the *balance* between the parameters of production, with an understanding of what can be managed most easily on site. Such an approach enables the construction manager to develop an effective production plan which

Table 3.1.1 Production parameter comparison: timber frame vs. brick/block

Production parameter	Timber frame	Brick/block
Time	Shorter envelope construction period – internal work can commence very shortly after the timber frame is constructed, potentially a matter of days. Must be factory ordered, so longer lead time to start on site. Not affected by cold weather.	Longer envelope construction period – internal work can only commence after the envelope has been built in its entirety, potentially a matter of weeks. Materials readily available so little lead time to start on site. Cannot be constructed in very cold weather; may lead to delays.
Cost	Approximately equal cost for the construction, but shorter on-site programme overall can reduce costs.	Approximately equal cost for the construction, but longer on-site programme overall may increase costs.
Quality	Factory produced, so high-quality product for the frame itself. However, the external skin will be site produced, so potentially quality issues if not managed correctly. Relatively small amounts of skilled labour in this trade. Potential issues with sound transfer if not correctly specified and constructed. Significant client changes not easy to incorporate.	Site produced, so potentially quality issues if not managed correctly. Relatively large amounts of skilled labour in this trade. Few issues with sound transfer due to the nature of the materials. Significant client changes relatively easy to incorporate.
Health and Safety	Scaffolding requirements – potential falls from height.	Scaffolding requirements – potential falls from height.
Environment	Very energy efficient if correctly specified and constructed. A sustainable construction material.	Very energy efficient if correctly specified and constructed. High level of embodied carbon.
Practical considerations	Site access may affect panel size that can be used (limited turning for wagons, etc.).	Site access only needed for pallets.

will meet the parameters, but also be implementable within the constraints of the construction site environment and the practicalities of construction work.

Communicating the plan

Once the planning process has been undertaken, and in order for the final production plan to be of any use, it should be easy to understand and communicate.

The most common way the plan is communicated is through a **programme** of work.

Although there are many different ways of illustrating the plan through a variety of programming techniques, the programme document should clearly show the following:

- The construction activities to be undertaken.
- The duration of these activities, and therefore the overall project duration.
- The sequence of these activities and the relationships between them.
- The resources needed to carry out those activities in the time indicated.

From this resourced programme cost reports can be generated, which show the rate of spend for the project over time. A site layout is also required to show how the site will be set up within the practical space in order to support the planned construction work. Quality should be 'built into' the programme – if the activities are managed and completed in the correct sequence and in the correct way, then they should easily be able to meet the quality requirements of the project.

The programme should be supplemented by comprehensive **method statements** and **risk assessments** to illustrate how these construction activities will be carried out in practice, and the necessary health and safety management of the work methods selected. An **Environmental Management Plan** should also be included, to ensure environmental parameters have also been considered and addressed.

The programme and this associated documentation should clearly demonstrate how all the production parameters have been considered and planned for, both individually and with an understanding of how they interrelate. The construction manager has effectively produced a 'to-do' list for the project, showing what needs to be done, how and when, in order to meet all the production parameters and deliver a successful project.

The detailed processes that are used to plan construction work, develop a construction programme and manage time, cost and quality on the project are discussed in more detail in Sections 3.3, 3.4 and 3.5. Establishing an effective site layout is examined in Section 3.2. For detailed discussions of health and safety management and environmental management see Sections 3.6 and 3.7 respectively.

Planning is only one half of the process however. If plans and programmes are produced, just pinned to the wall and forgotten about, they become useless. What is also essential to construction production management is *control*.

3.1.5 Production control

Control is Fayol's final principle of management, and is the checking of actual performance against the plan. Three steps are involved in production control:

- Production of a clear plan.
- Comparison of actual events with the plan.
- Taking corrective action if the actual events are not meeting the plan, or change the plan.

Control may be undertaken in many different ways. Site visits and inspections form a key part of the control aspect of construction management. Simply being on site and seeing what is going on enables the construction manager to make a constant comparison against the plan, and on small sites this may be enough.

On larger sites, formal reporting channels are often in place, with the construction manager producing a detailed report of actual events, often termed a **progress report**, to send back to head office. Such reports can be produced on a weekly or monthly basis, depending on the volume of construction activity being undertaken and how close the project is to completion. These reports may also contain photographs of the site as the works are progressing, marked-up drawings, and progressed or marked-up programmes showing the actual position of the project against the plan.

Key questions to be asked are:

- Are we on time? (often referred to as 'on-programme')
- Are we on budget?
- Are we meeting quality requirements?
- Are we meeting health and safety targets?
- Are we managing our environmental impact?

How these questions are answered is explored in more detail in the following sections, where the planning and control of each production parameter is examined in more detail.

For these questions to be answerable, however, the initial plan must itself be both clear and achievable.

If any of the above questions receives an answer of 'no', then corrective action must be taken to bring the project back in line with the plan. This may mean either a revision to the plan while still remaining within the original production parameters, or a more fundamental change that can affect the parameters in operation. For example, if work is falling behind in one area, extra resources can be brought in to speed up the rate of production. Obviously this will have an impact on the cost of the project at that stage, but the contractor may find this a better investment than not completing the whole project on time.

3.1.6 Production planning and control: summary

Construction managers must plan in order to ensure that they are able to meet all the elements of production necessary for their project, within the boundaries of practical construction considerations. They must also plan in such a way that they are able to control the work once it has commenced.

This means that the construction manager must think about their project as a *whole* when planning and controlling it. Just focusing on one element, for example, cost, may mean that less attention is paid to time or environmental concerns, and although the project may complete under budget it may overrun its deadline and not meet the environmental requirements of the client, which can be costly in both **liquidated and ascertained damages (LADs)** implemented under the contract, as well as loss of reputation.

The construction manager should therefore ensure that the work keeps to the plan wherever possible, but change can and often does happen, and so the plan needs to be flexible enough to allow alteration as the works progress.

For example, if the client makes a change to the design, this could affect the programme of work in terms of activity and sequence, which will require a revision to the plan. Under the contract, any delay or increased cost will be paid for by the client and so the construction manager's role is to incorporate the changes as efficiently and smoothly as possible, given all other production parameters.

Alternatively, poor-quality work could require knocking down and rebuilding, so the construction manager must decide whether to stop work to rebuild this section before continuing with the programme as planned, or reorganise the planned sequence around the faulty work to avoid delaying the entire project. In this example, the fault could lie with the construction manager if they have not kept a close eye on quality management, and so any costs may have to be borne by the contractor. The construction manager must ensure that there is minimal impact on the project as a whole, and any other production parameters, as a result of poor workmanship. They should also revise their quality management system to make sure it does not happen again.

More fundamentally, problems with the weather and ground can also bring surprises and changes to the project. If the weather is much worse than expected, cranes may be 'winded-off' and progress limited or even halted for long periods of time. Unexpected ground conditions can also cause delays, and the project may need to be redesigned to meet the new loading requirements, or archaeological findings may be revealed that can stop work for a period of time while the findings are assessed by the local council's archaeological team.

Production planning and control is therefore an ongoing process that forms a significant part of construction managers' daily work. Construction managers must spend the majority of their time on the site rather than in the office. They need to check that work is going according to plan, that change is being managed correctly, and control measures are being implemented where required. This means keeping a close eye on time, cost, quality, health and safety, and environmental management, all at the same time.

In order to make this possible, and to help construction management in practice, various systems and tools have been developed to ensure that nothing is missed. The production parameters are now considered in the following sections, which examine how they are both planned and controlled on site on a daily basis.

Discussion point comments

 3.1

Discussion Point

What other relationships can you see between these five elements?
What are the different results of optimising the various elements?

The relationships among the five parameters of production can be very complicated.

Table 3.1.2 suggests a possible effect which the optimisation of each may have on the others.

In Table 3.1.2 pros and cons to the optimisation of all the parameters of production are identified, and in some cases a pro *and* a con may be identified.

However, it will also be seen that through optimising quality there is the potential to improve all the other parameters of production. This shows the importance of quality management within the construction industry and why it should be at the very heart of good construction management.

Please see Table 3.1.3.

 3.2

Discussion Point

How could the findings of the site visit affect production planning?
How could they affect the different parameters of production?

Bad planning is the result of bad information. If all the available documentation has not been fully examined, or if a site visit was not undertaken in depth, managers may end up making wrong decisions which cannot actually be implemented in practice.

The best way to prevent this from happening is to develop a management system (which should be part of a wider quality assurance system if the company has one in place) to make sure that all aspects are considered at the tender stage. A detailed standard checklist should be used when making a site visit to make sure that everything has been thought about and checked when on the site rather than letting people make up their own checklists, as ideas of what is important can differ greatly from person to person.

 3.3

Discussion Point

What could lead to bad planning by the contractor?
What would be the best way to prevent this from happening again?

Table 3.1.2 Relationships between the five parameters of production

Time	Cost	Quality	Health and safety	Environment
Optimised (quick)	May increase to accelerate the programme	May reduce as less time for activities	May be more incidents as less time to plan work safely	May be more incidents as less time to plan work safely
Work may take longer as less resources can be applied	**Optimised (cheap)**	Cheap workers may be poor quality, and may rush to complete the work in the available budget	May be more incidents as less investment in safe working practices	May be more incidents as less investment in environmentally sound working practices
Work may take longer to produce a quality product May take less time as quality processes improve planning	May be more expensive for quality materials and workers May be cheaper as quality processes increase efficiency, less waste	**Optimised (high quality)**	May be fewer incidents as a quality product takes time and good access, which is often safe access	May be fewer incidents as a quality product takes time and good work practices May improve environmental impact if products can be selected on environmental criteria
Work may take longer to carry out safely, but better planning and access may make it more efficient	May be more expensive to use the safest methods and invest in health management	May improve quality as better cared-for workforce	**Optimised (safe and healthy)**	May be fewer incidents as better work practices and happier workforce
May take longer to make sure environmental controls are in place	May cost more to implement high levels of environmental management	May affect quality as sustainable products and practices are used – for better or for worse	May reduce incidents as green approach also includes workforce care	**Optimised (green)**

Table 3.1.3 Findings and potential problems arising from the site visit

Check	Potential findings	Problems?
Site access and turning	Access too small or tight for standard wagons.	Split loads can be more expensive and inefficient, adding to cost. Fixed wheel wagons rather than articulated wagons are needed.
	Access too tight for some construction methods.	Large panel systems or frames will need delivery wagons of a certain size; if they cannot be accommodated another construction method must be selected, affecting all parameters.
Potential security issues	Poor neighbourhood. Very secluded area.	Cost of 24-hour security added to the project. Cost of secure containers and plant stores.
Who are the neighbours?	Constrained site with many neighbours.	Environmental issues with noise, vibration, dust once works begin.
	A school next to the site.	May restrict site operational and delivery hours, affecting time and cost.
		May be issues with parking.
Existing structures on the site	Structures on the site not noted on the project drawings.	May impact upon foundation construction and add time and cost.
		May change the proposed design, affecting all parameters of production.

Topography	Sloping sites, poor ground, fill.	May impact upon foundation construction and add time and cost. May be contaminated and require treatment, again adding to time and cost as well as environmental concerns.
	Potential ecological habitats.	May impact upon time and cost in management of local ecology, changes to design to accommodate.
	Potential archaeological issues.	May impact upon time as inspections are carried out to permit work to continue.
Services	Underground services not shown on the utilities information provided for the project.	May change the proposed design, affecting all parameters of production. May affect the foundation design. May be legal covenants on the services that will require access.
	Overhead services not shown on the utilities information provided for the project.	May affect crane use and therefore proposed method of construction (e.g. large panel system) and all production parameters.
Other provisions	No parking in the local area.	Potential cost of providing parking on site, or managing drop-off areas for tools and materials.
	Watercourse near the site not shown on the project information.	Environmental concerns regarding potential pollution management and increased cost and time to the project.

3.2 Site management

3.2.1 Introduction

Site management is focused on the planning, setting out and ongoing control of the construction site. The overall aim of site management is to ensure that the optimum site layout is established at the very beginning of the project, and all necessary management provisions are in place to support effective and efficient production management. Site management must also be flexible enough to adapt to any changes that will happen during the construction process, including the construction of the final project itself.

Some site management provisions are required by legislation, which sets minimum standards for UK sites. For example, the necessary **welfare** for the site is prescribed, and includes toilets and places to eat, as well as other health and safety and environmental considerations.

The cost of site management is built into a project through what are known as **preliminary items** or 'prelims'. These are elements of the work that support its construction, but are not part of the completed project; for example, scaffolding, temporary roads and site fencing or hoarding. Prelims also include the salaries for the site team, as well as any other necessary provisions to enable the work to be carried out effectively and efficiently. Preliminary items are included in the **tender** for a project, and therefore early planning allows accurate budgets to be set within the tender submission. This means that money should be available within the contract for good site management once the project begins. If construction managers are involved from the pre-tender stage, they are able to have full input into these requirements and can make sure a suitable budget is set for effective and efficient site management of the project.

Figure 3.2.1 Construction site: Keep out!

An efficiently planned, laid-out and well-managed site will enable the maximum amount of work to be undertaken without unnecessary delays occurring due to poor distribution of plant and materials, or poor access or loading provision. Good site management makes a site effective and efficient, saving both time and money for the project. A good site layout will also create positive attitudes towards the project as a whole within the workforce and the general public.

Without planning, sites can become established on an almost ad-hoc basis as cabins and stores are delivered and simply placed wherever is convenient at the time – this is how poor site layouts come about. Messy and inefficient site layouts can result in poor quality of work as operatives become frustrated, cause a lack of respect for health and safety and environmental concerns, and make a poor impression on the general public and the workforce about the contractor in charge.

This section looks first at the legislative requirements for site management. Site layouts and their key elements, including how to position them on the site, are then discussed. The effect which construction work can have on the surrounding environment is considered, and the need to challenge and change the common idea that construction industry projects are bad neighbours, and what can be done to combat this. Finally, the ongoing process of site management is examined, and how continued planning and control are needed to make site management an effective part of good production management.

3.2.2 Legislative requirements

Good site management is an integral part of good health and safety and environmental management, and is affected by legislative requirements in both of these areas.

The Health and Safety at Work etc. Act 1974 specifically notes under Section 2(2)d that employers must provide a safe place of work, maintain it in a safe and healthy condition, and provide safe access and egress to it. The Act also requires adequate facilities and welfare provisions to be in place on sites.

These requirements are also included within the Construction (Design and Management) Regulations 2007, and are set out in much more detail and with specific reference to the construction industry. Section 4 of the Regulations details the requirements, which include, among others, consideration of the following:

- Safe places of work.
- Good order and site security.
- Traffic routes.
- Emergency routes and exits.
- Fire detection and fire-fighting.
- Lighting.

The Regulations also contain within Schedule 2 the requirements for welfare provision on construction sites. Details of how to meet the requirements, depending on the size of the site, may be found in the Approved Code of Practice (ACoP) developed to support the CDM Regulations. This is available free of charge from the HSE and available to download from its website.

Environmental concerns are addressed through the Environmental Protection Act 1990, which details several statutory nuisances that must be reduced where possible. These include the following:

- fumes;
- dust;
- noise;
- vibration.

All of these nuisances can be produced in one way or another by construction activity, but can be reduced and even eliminated with good site management.

Other environmental legislation is specifically focused on waste management and is covered by a duty of care and the Waste Regulations 2011. These set out the requirements for waste management and control on sites.

Environmental legislation also penalises pollution, and within construction activities pollution can be a direct result of poor site management. In the UK the law states that the polluter pays, and so incidents can be costly in both time and money.

Discussion Point

How could pollution be caused on sites?
What may be done to mitigate against this potential environmental impact?

There is also the potential for polluting contaminants to reach surface and groundwater sources with serious repercussions. More considerations of environmental management with relation to construction activities may be found on the Environment Agency website and in Section 3.7.

Other issues that can affect site management and have potential legislative consequences are managed and implemented by local councils; these include the following:

- Maintenance of public rights of way.
- Maintenance of pavements and verges.
- Maintenance of highways.

Good site management must make sure that there is a plan in place to avoid any issues in these areas. For example, delivery vehicles or workforce parking on local roads can cause damage to footpaths and highways, especially if the roads are not constructed for that weight of vehicle. **Dilapidation surveys** should be undertaken to show where roads or pathways are in poor repair before work starts, to ensure that contractors are only liable for damage they may cause, not damage that was already there. Remedial works may need to be factored into the project costs, but it is better to avoid causing damage in the first place and to plan a system that enables close management of deliveries to eliminate waiting, for example, using **just in time** deliveries to make sure materials arrive as they are needed in the work.

Discussion Point

How would you know if there was a right of way on the site? What could you do about it during the project?

All of these different legislative requirements must be considered when planning the site management. Although these requirements should be included in the initial plan, the development of the project should also be considered to make it as easy as possible for people to continue to comply with legislation once the work begins.

3.2.3 Site layout

Once legislative requirements have been established, the construction manager should develop a site layout, which often takes the form of a marked-up drawing. Site layouts can also be developed using BIM software and applied to the BIM model, if it is being used on the project. This site layout communicates the site management plans. It should contain the following as a minimum:

- Location and specification for the site boundary.
- Site identification signage.
- Site security provision.
- Site access gates (location and type).
- Traffic/pedestrian management within the site.
- On-site parking provision (if any).
- Locations of temporary services.
- Building footprint.
- Locations and types of access to the building (e.g. scaffolding, loading platforms).
- Emergency escape routes and fire-fighting provision.
- Site welfare location and provision.
- Site office location and provision.
- Material stores.
- Plant management.
- Waste management provision.

All of the above should be carefully thought about and planned before the project even starts on site.

An example site layout for a six-plot housing development is shown in Figure 3.2.2. We will review the layout at the end of this section to see why the site has been set out as shown.

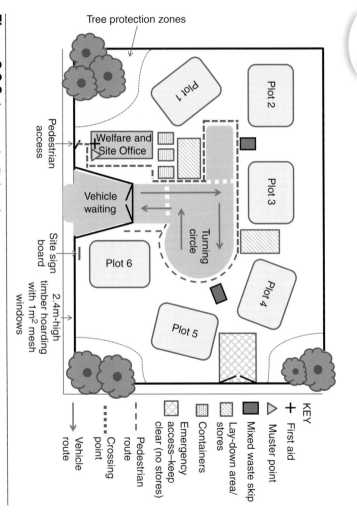

Figure 3.2.2 An example site layout

The site layout also needs to be flexible. On large projects there may be several site layouts for the different phases of the project as the work develops and the project is constructed. For example, **enabling works**, carried out at the very beginning of a project, will require a different site layout to the finishing works in terms of provisions. In fact the enabling works can incorporate works to develop the main site layout such as providing **hardstandings** for site cabins or final drainage connections.

Site layout should be reviewed alongside the programme of works. The construction manager must consider how the daily changes on site and progression of the programme will affect the site management, and make adjustments accordingly to ensure that the optimum site layout is maintained. This is discussed in more detail in Section 3.2.5.

Site boundary

First, it is essential to make sure that the site boundary matches exactly the legal land boundary for the project, and this will include any land that must be crossed for access. If the land and necessary access is not under the client's ownership there may be problems with allegations of trespass, and mistakes can be very costly in terms of both time and money while legalities are being sorted out.

Under CDM Regulation 27, site boundaries must be clearly defined and fenced off to prevent unauthorised access. There are several different methods that can achieve this requirement. Boundaries may be constructed using open-mesh panel fencing, solid metal panel fencing or a solid timber hoarding.

Solid timber **hoardings** are the most robust, using timber posts set into concrete along the boundary, to which the plywood panels that form the hoarding are securely fixed. Some clients specify timber hoarding to be decorated in their corporate colours so that members of the public know this is their project. Some large contractors also have a corporate colour scheme for their hoardings, again to show that they are in charge of the site. Mesh fencing boundaries, often known by the brand name 'Heras', are more flexible and can be moved; however, they are not as robust and offer no protection against dust and noise emissions from the site unless they are sheeted. Solid panel fencing is also flexible and offers more protection against dust and noise than an open mesh panel. Examples may be seen in Figures 3.2.3, 3.2.4 and 3.2.5.

Discussion Point

3.6

What pros and cons could there be with the different boundary methods?

What production parameters should be considered?

In some cases the permanent site boundary (e.g. a high brickwork boundary wall) can be constructed early in the project which removes the need for a temporary site boundary to be installed.

The site boundary forms the public face of the site and so must be kept in good order. Often, on large projects, mesh 'windows' are included in a timber hoarding so that people can see what is happening on the site to satisfy their curiosity, but it can also be a deterrent to thieves as they could be spotted breaking into the site from the street.

Figure 3.2.3 A solid timber hoarding site boundary

Figure 3.2.4 An open mesh fencing site boundary

Figure 3.2.5 A solid panel fencing site boundary

Site identification signage

The site boundary also provides a good fixing point for the site identification signage, and any other key information such as the main project participants, the type of project under construction and the site safety requirements. Examples of this safety signage are given in Figures 3.2.3 and 3.2.4, as required by CDM Regulation 27.

Site security

The site boundary forms an essential aspect of site security, preventing people or children from gaining unauthorised access to and from the site, and guarding against theft and vandalism.

Security may be provided at all access points to the site, with a security cabin and security guards to check people, wagons and materials on and off the site if the budget permits.

On large projects, the security cabin often forms the main entrance to the site, and ID cards, facial recognition or fingerprint technologies are used to monitor who is on the site at any given time. The workforce cannot access the site unless they have the appropriate card or log-in. Such a system is shown in Figure 3.2.6.

On smaller projects, the luxury of site security guards may not be possible within the budget. They will not be as busy as large sites, with fewer deliveries and a smaller workforce, and so a full-time security guard may not be needed. Instead, careful positioning of the site offices next to the site access, and stacked if possible for better visibility, allows site managers to keep an eye on the site access and

Figure 3.2.6 Turnstile entrance to a site

Figure 3.2.7 Site offices positioned to aid site security

enables entry and exit through the gates when needed, as is shown in Figure 3.2.7.

But on both large and small sites security must be considered when work has stopped for the day. It is the responsibility of the contractor to ensure that children cannot access the sites at evenings and weekends. Construction sites are also vulnerable to theft and vandalism. On large sites, 24-hour security in the form of on-site security guards or regular inspections will probably have been accounted for in the preliminaries, and can prove a strong deterrent. CCTV systems may also be used, although they tend to be more reactive than proactive in preventing incidents.

On smaller sites without such budgets, security may simply be a robust hoarding and the securing of plant, materials and the project itself each night. In these cases, developing good relationships with the local neighbours can be a good idea, as they are likely to be the first to hear or see anything happening out of hours on the site, and could contact the police or contractor's representative to raise the alarm.

Site access gates (size and type)

A key consideration is the size and location of the site access gates for both traffic and pedestrians coming on to the site to work. Separate gates are required, and should be clearly marked to ensure that people are not using the traffic gates for access. An example of a segregated entrance may be seen in Figure 3.2.8.

Access for traffic to the site will be dictated by the site itself, and how constrained and restricted in terms of space it is likely to be. Ideally,

Figure 3.2.8 A segregated pedestrian entrance to a site

two separate gates will allow a one-way system for deliveries, removing the need for vehicles turning within the site, although this can affect security provision which will need to be doubled up.

The access provision must also take into account the size of the delivery vehicles and their manoeuvrability both within the site and immediately outside it. For example, if the access gate is positioned too near a road junction, delivery vehicles may not be able to turn into the gate easily or safely. Delivery vehicles stopping on the main highway while they wait to gain access to the site should also be considered. This can cause traffic obstructions and unnecessary hazards in the roadway. Where possible, gates should be set back into the site so that vehicles can pull into a 'lay-by', still outside the gate but off the highway, while they wait for permission to enter and unload.

Discussion Point

How could site access affect and potentially be a hazard to the public?
What measures could you put in place to help protect them?

Traffic/pedestrian management within the site

Traffic management also forms a vital part of safety management on the site and is legislated for under CDM Regulation 36 due to the health and safety risks; being hit by a moving vehicle is a common cause of construction injuries, although this is reducing in frequency as more attention is paid to traffic and pedestrian management on sites.

Traffic management may require the construction of temporary roads within the site to ensure that vehicles can easily manoeuvre within the site and do not become stuck in mud. If there are to be permanent roads on the project, these can be partially constructed during the enabling works and used for access. Often, the road is built to the base course of tarmac, so any necessary remedial works can be carried out before the final top course is laid and the project handed over.

Pedestrian walkways should be clearly defined and maintained through the use of physical barriers and signs, keeping people away from the designated traffic routes. On the site shown in Figure 3.2.9, segregated walkways have been established through the use of barriers. The gates show a pedestrian crossing over the materials storage area and give a signal that vehicles may be manoeuvring in that area.

Vehicle routes must also be clearly set out, with areas allocated for loading and unloading. Where a one-way system cannot be implemented on the site, vehicle routes need to consider the turning requirements for different-sized vehicles to make sure that they can safely turn in the available space. Site speed limits should also be set and enforced to prevent accidents and minimise raising dust.

Traffic and pedestrian management should be clearly shown on the site layout, and a formal traffic management plan (TMP) produced to illustrate not only how it has been addressed but also how it will be monitored and maintained. This traffic management plan should form

Figure 3.2.9 Segregated walkways on a large site

part of the Construction Phase Health and Safety Plan as required by the CDM Regulations.

Management control of the traffic management system is essential as barriers can be easily moved.

Discussion Point

Why might traffic or pedestrian barriers be moved?
How would you manage this as the construction manager?

On-site parking provision (if any)

The parking a site workforce needs can be one of the biggest neighbourhood problems for construction projects, as cars and vans mean that local residents cannot park themselves or easily gain access to their homes. If on-site parking is possible, it should be provided to reduce any problems. However, in many cases it simply cannot be fitted into the space available and the workforce must be encouraged to be as considerate as possible and not to cause the project any undue issues through poor behaviour.

Locations of temporary services

Temporary services that may be required for the project are as follows. The

Water supply: Water is required on site for several reasons. The welfare facilities, if plumbed in, will require hot and cold running water for the WCs, urinal stalls, wash-basins and shower units. Some construction activities also require water for mixing materials, such as plastering, and water may also be needed for damping down work areas to avoid dust and vehicle wash-down.

Electricity supply: Electricity will also be needed for the welfare and office facilities for the provision of heat and light, if they are not to be run from a generator. Transformer positions will also be required on site for non-battery-powered plant to drop the supply from 240V used in the cabins to the less dangerous 110V used on sites. External lighting may also be needed for winter working or site security. A temporary supply for the site must be connected by a fully qualified electrician and securely controlled according to CDM Regulation 34. In some cases a 415V supply is needed, for example, for powering tower cranes and testing lift installations, and again these connections must be installed by a fully qualified electrician and made secure. Such an installation may be seen in Figure 3.2.10.

Drainage: Although chemical toilets may be used to avoid the need for an on-site drainage connection, or waste stored in tanks and pumped out of welfare provisions, where possible it is much better to connect to the main sewer for health and safety reasons. Surface water drainage must also be controlled and measures put in place to make sure no pollution enters the system from the site activities. Early installation of the permanent connection to the surface water system can help site management control water on the site and ensure that it drains away quickly, preventing the site from becoming waterlogged and making working conditions difficult.

Gas supply: Gas is rarely used on sites, but if gas is to be used for heating cabins the necessary health and safety legislation and procedures apply for its use. One benefit of gas is that it is silent in its provision of heat and light, and special gas-powered welfare cabins may be used where noise is a serious issue.

Telephones/Internet: With wireless technology now widely available, this connection is not as essential as it once was and may not even be necessary. However, hard connections to the telephone and Internet may be put in place to ensure a reliable working environment at all times.

Figure 3.2.10 A secure electrical services provision

Building footprint

The site layout must clearly show the building footprint to enable the accurate positioning of all other site management considerations. Although 'footprint' technically only refers to those parts of the building that are in contact with the ground, any overhangs or cantilevers must also be noted on the layout, as these can restrict access and vehicles manoeuvring within the site.

Locations and types of temporary access to the building

Temporary works or access requirements needed as the construction progresses can be accurately positioned on the site layout around the building footprint.

For example, scaffolding may be required to the perimeter of the building and will need good level ground to build from. Alternatively, if access is to be provided by mobile elevated working platforms (**MEWPs**) or **cherry pickers**, firm and level ground will be needed for safe working. This means that temporary hard surfaces or hardstandings must be constructed around the building and kept clear for access machinery use only.

Other access provision may be required in the form of loading bays or hoists in the side of the building **envelope** so that materials can be loaded easily to the upper floors. This means that a working space on the ground is needed to allow forklift trucks to safely load and remove materials as needed. Timing this type of access should be carefully planned. For example, a loading bay on the side of a large block of flats will need to be closed up at some point in the programme to enable the external and internal finishes to be completed in that area, but this may be before work on the flooring is due to start on site. Vinyl flooring materials are both large and heavy, and should not be manhandled if possible. It is the role of the construction manager in their planning and management of the site to think about this in advance. They should pre-order the materials in good time and bring them into the building before loading bays are removed. This will make sure that they are brought safely into the building without unnecessary risk to the flooring operatives, damage to the materials themselves or the parts of the building already completed.

The construction methods agreed in the initial planning stages of the project will dictate the need for any such access and at what stage in the project. This access should be included in the project programme as well as on all revisions of the site layout plan. Temporary access may affect the working space of the site, and the construction manager must make sure that the location and provision of such access:

- enables the construction works to be undertaken as planned;
- does not impede vehicle or pedestrian access;
- is not affected by the provision of material stores or lay-down areas;
- allows safe and efficient access for loading materials into the building;
- can be maintained for as long as it is needed during the construction work.

Emergency escape routes and fire-fighting provision

Although a construction site is not a completed building, emergency escape routes still need to be set out and managed to ensure that they are kept clear. This is legislated for in CMD Regulation 40.

CDM Regulation 41 notes that construction sites can be vulnerable to fires, and suitable measures must be put in place to protect people on the site from the risk of fire.

3

3.9 Discussion Point

What construction activities might cause a fire on site?
How do you think this could be managed?

Different fire extinguishers are used to fight different types of fires, and are colour coded to indicate their use. Fire points should be established at key locations around the project, and the most suitable extinguishers for the surrounding environment made easily available in case of emergency.

A muster point is also needed for the workforce to gather if an emergency alarm is given. This should be in a safe place, away from vehicles and the construction works, and should be clearly indicated on the site layout.

3.10 Discussion Point

How would you show the emergency escape routes on site?
How would you make sure they were kept clear?

Site welfare location and provision

Site welfare is legislated for in the CDM Regulations 2007 and requires provision of the following:

* toilets;
* washing facilities (with hot and cold running water);
* provision of drinking and hot water;
* changing rooms and lockers;
* drying rooms to dry clothes;
* showers;
* facilities for rest.

Purpose-built site welfare cabins are readily available for hire, and can even contain all of the above within a single cabin. Welfare requirements depend on the size of the workforce on site at any given time, which dictates the number of toilets needed, the size of the changing rooms, and the number of lockers. For example, on a site with a maximum of 50 workers at any one time, three toilets and three wash-basins will be required (HSE 2014a).

Cabins are available in many different sizes to meet the necessary requirements in a variety of different combinations.

The site shown in Figure 3.2.11 is currently carrying out enabling works and there are only a few people on the site. This combined cabin is split into three, and contains the office, toilet and canteen all within one unit.

On larger sites, cabins may have just one use and can be stacked; for example, welfare and office cabins are often stacked on top of each

Figure 3.2.11 A combined welfare/office setup

other for ease of service connections and also to save space on the site. On very constrained sites, such as in city centres, the cabins can be stacked above pedestrian walkways or access points to the site, as is shown in Figure 3.2.12.

The location of the site welfare needs to be carefully considered to make it convenient and practical for the workforce to use. Welfare cabins are usually positioned by the site entrance so that workers can easily change into their work clothes when they arrive on site, and leave anything they don't need on site in their lockers. This also means that workers can shower and change before they leave the site. On large sites smaller on-site welfare units may be needed, and on high-rise projects installing toilets on the upper levels of the building is also a good idea to save people walking all the way down to the site entrance to use them.

Site office location and provision

The size of the site offices will be dictated by the size of the project and the number of staff working on the site. Site offices are also contained within cabins, and come in a variety of sizes and layouts.

The location of the site offices can be critical in maintaining control of the site. On small sites, the offices should be placed by the site access gates in order to monitor anyone entering and leaving the site. It is also a good idea to locate the offices near or facing the construction works themselves so that staff can easily keep an eye on the work as it progresses.

Figure 3.2.12 Cabins stacked above the site entrance

Material stores

Careful planning of materials storage is essential to ensure minimisation of waste, and the location of the stores should be clearly communicated through the site layout. Material delivery should be made on a just-in-time (JIT) basis to ensure that excess materials are not being stored on site, making them vulnerable to theft and damage. More about JIT management may be found in Part 4.

Positioning the stores within the site layout must acknowledge the fact that materials will need to be delivered into the stores and taken out again by the workforce to use in the construction. This means that the stores must be easy to access for everyone. For example, light materials can be unloaded and carried by hand but heavier materials may require dollies (small wheeled platforms), pallet trucks or the use of the forklift to move them from the delivery vehicle into the stores, and from there into the building. This means that the position of the stores must be mindful of the traffic and pedestrian management plan for the site.

The location of the stores should also make returning materials unused in the day's work the easy option. This reduces buildup of materials on site and reduces waste. Any spare materials taken on to site that day may be returned to the stores rather than to the skips to minimise site waste.

The type of store needed depends on the materials to be stored. Larger materials cannot be placed inside containers and instead will require designated **lay-down areas**. The same considerations apply in terms of ease of access, and these external stores are exposed to the weather and should be on a dry, firm hardstanding to avoid waste through damage or contamination from the ground. Other materials should be safely locked away; for example, copper pipe used in plumbing installations is relatively valuable and prone to theft, and so should be stored within lockable containers.

In Figure 3.2.13, a rack has been built to store the materials needed for the first fix mechanical and electrical works on this large construction project, keeping them tidy and easy to access.

As the works progress, storage areas inside the building may be needed for certain materials, such as flooring or final fix joinery items. Once materials are secured inside the building this should be taken advantage of as it means that materials are only handled once: from the delivery vehicle into the building, rather than from the delivery vehicle to an external store and then inside the building, making it a more efficient process. Temporary lock-ups may be provided – designated rooms with lockable doors for different trades to store their materials – but care must be taken to protect any finished works within the rooms from damage.

Avoiding repeated or **double handling** is a key part of materials management on site, as the more materials are handled, the more potential there is for damage and waste. Incorrect storage of materials can also lead to damage; for example, some materials are vulnerable to weather and some need protecting.

The materials storage schedule shown in Table 3.2.1 on page 102 provides more guidance on storage requirements for common domestic

Figure 3.2.13 Site built storage rack

Table 3.2.1 Materials storage schedule

Material	Storage requirements	Things to note
Drainage	Keep in crates or on wrapped pallets on firm, clean ground.	Vulnerable to breakages. Can sink into mud.
Reinforcement	Designated lay-down on a hardstanding. Mat reinforcement to be laid flat. Tying wire should be stored in a locked container.	Can sink into mud. Vulnerable to contamination and distortion.
Scaffold components	Designated lay-down on firm, clean hardstanding or purpose-built racking.	Clips easily sink into mud. Clips are sometimes stored in diesel which is an environmental hazard. Scaffold boards are vulnerable to being put to other uses on site.
Bricks/blocks	Keep on wrapped or banded pallets on firm clean ground or hardstanding.	Split pallets can become easily damaged. Can sink into mud.
Mortar	If scope allows, a dry or wet mix mortar silo can be installed on the site.	Site mix mortar cement must be protected from weather or it will solidify.
Timber	Structural timber must be kept off the ground and stored horizontally. Roof trusses must be stored vertically on racks. Finishing timber must be stored internally on horizontal racking, but maintain air flow.	A temporary scaffold 'shelter' may be used for storage. Can deflect and weaken with incorrect storage and handling. Must be kept dry.
Roof finishes	Keep in wrapped pallets on firm, clean ground. If slates are being used/reused stack carefully on hardstanding.	Vulnerable to breakage.
Electrical components	A locked container.	Vulnerable to theft.
Plumbing components	A locked container.	Vulnerable to theft. Ceramics are vulnerable to breakage.
Plaster	Must be kept dry.	Vulnerable to theft.
Paint	A locked but ventilated container.	Extremes of temperature must be avoided.

construction products, although where possible advice from manufacturers or suppliers should always be sought.

Construction managers must not only plan the location and access to and from the stores, but also ensure good management of the stores once they are in everyday use, to keep them clean, tidy and safe.

Plant management

Plant should be stored in a designated compound which must be secure but also allow for easy access to the plant as and when it is required. This compound should be clearly identified on the site layout. On smaller projects plant may have to be left within the main site boundary, but can be protected by shutters, as is shown in Figure 3.2.14, or fitted with tracking systems in case of theft.

Small plant should be stored securely in a lockable designated metal container within the main plant or materials storage area.

Fuel for plant should be kept in a designated area in a bunded fuel bowser. Any refuelling should be carried out over a drip tray to prevent spillages, and the designated refuelling point shown on the site layout.

On large projects, any tower cranes required must be included in the site layout plan, with allowance for their erection and dismantling. The site layout should also include the coverage of the tower crane's jib in order to ensure access to the materials stores and delivery vehicle unloading areas as necessary, and that workers do not over-sail neighbouring properties without permission.

Figure 3.2.14 Shutter protection on a road roller

Waste management provision

Like material stores which should be planned for easy movement of materials from the stores to the construction workface, waste management provision for the site should be planned to enable easy movement of waste in reverse, away from the construction workface and into skips.

Positioning skips near the exit from the building next to the pedestrian walkway can encourage people to use them and keep their work area clean and tidy. Access for forklifts in unloading bins may also be needed, and the skip wagon will certainly need access to replace the skips easily and safely when they need changing.

3.11

Discussion Point

Skips need access for people, machinery and vehicles.
How would you manage this area to keep it safe?

Duty of Care legislation means that waste on site must be properly managed, and skips provide the most practical solution for this, as well as a separate hazardous waste skip or drum. Many sites set up waste segregation points, with separate skips for timber, metal and general construction waste. One benefit of segregated skips on site is that the construction manager can see what materials are being wasted and which work activities are producing them, and so may need closer management. But many waste carriers will now collect mixed skips from site and sort at their own depots for recycling, and this can be a more economical approach. It is the responsibility of the site to ensure that the waste carrier is authorised to carry waste, and is properly registered. All waste must be sent off site with a completed waste transfer note, which states the type of waste produced.

Construction does produce some specialist waste that requires more complicated management. For example, plasterboard and gypsum products cannot be sent to landfill mixed with other waste. Some waste carriers will be able to sort plasterboard from mixed skips at a cost, but it may be easier to segregate this material on site.

Other materials are classed as **hazardous waste** and *must* be segregated from other waste, and may only be removed by a carrier with a hazardous waste permit. Hazardous waste includes some paints, silicone sealants and mastics. Hazardous waste should be stored in a locked skip or drum to keep the waste secure and prevent water from getting in, which could result in pollution.

More information about waste management provision may be found on the Environment Agency website. This aspect of environmental management is also discussed in more detail in Section 3.7.

An example site layout

Looking back at the site layout shown at the beginning of this section, the reasons behind the positioning of the different elements may now be better understood.

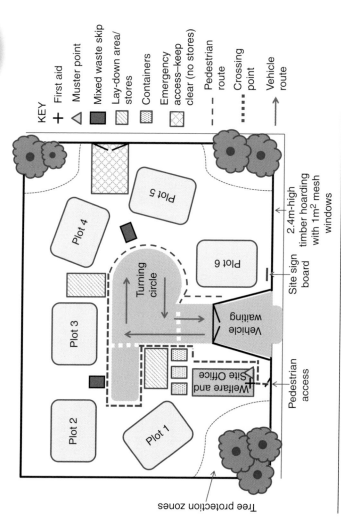

KEY

+ First aid
◁ Muster point
▨ Mixed waste skip
▨ Lay-down area/ stores
▨ Containers
▨ Emergency access–keep clear (no stores)
- - - Pedestrian route
···· Crossing point
→ Vehicle route

Plot 5

Plot 4

Plot 3

Plot 2

Plot 1

Turning circle

Plot 6

Vehicle waiting

Welfare and Site Office

Pedestrian access

Site sign board

2.4m-high timber hoarding with 1m² mesh windows

Tree protection zones

Figure 3.2.2 An example site layout

All the key elements have been considered. The site boundary is clearly marked as are the footprints of the houses, meaning that a detailed site layout can be planned. The hoarding has been specified, and set back at the gate to prevent vehicles from waiting on the road, and a splay included to allow visibility for drivers when turning out of the site. A set of emergency gates and access have also been provided in case the main gate cannot be used in an emergency.

The site is too constrained for a one-way access, so the final road layout has been used to incorporate a turning circle. Pedestrian access has been segregated from traffic, and walking routes inside the site are clearly marked, including crossing points.

The site office has been placed next to the gates to monitor who enters and leaves the site. The offices are stacked above the welfare cabins to give the construction manager a good view of the gates and the whole project. The offices also contain the first-aid point for the site. The muster point in case of emergency is located near the pedestrian gate and away from the vehicle route.

Stores have been clearly marked, including both containers and lay-down areas for pallets. Two skips have been provided to make sure one is always close at hand to encourage their use. The existing trees on the site have also been protected with clearly marked zones and fencing to make sure they are not damaged during the project.

The site layout should be clear and easy to understand, and include all necessary details to allow someone who has just started on site to quickly find out how to get around the site safely, as well as to plan and carry out their own work as efficiently as possible.

3.2.4 Being a good neighbour

The construction industry has not always had a good reputation as a neighbour: we start work very early in the morning, we are noisy, we create dust, we make surrounding buildings shake, we track mud on to roads that either becomes a dangerous slip hazard or dries and makes even more dust, we bring extra traffic and parking, we cut down trees and we build on green spaces.

But although the industry can also bring a lot of good to an area, such as local employment, regeneration through our finished products, and increased trade in local shops and cafés, these benefits are often overshadowed by what people see as the negatives.

Good site management can do a lot to help change the image people have of construction, and to promote a company as a better construction contractor than most.

3.12

Discussion Point

What benefits could there be to developing good relationships with the neighbours?

There is also the suggestion that all companies should give something back to the areas they work in through **corporate social responsibility**. Often construction companies help organise community activities, mainly involving building or maintaining some project, supporting similar local events or visiting schools to make sure the local children know the dangers of playing on the site.

3.13

Discussion Point

Do you think construction companies should invest in the local communities they work in? Why?

The Considerate Constructor's Scheme

The Considerate Constructor's Scheme (CCS) was launched in 1997 as a way to try to combat the poor neighbour image of the industry. Sites register voluntarily with the scheme and are assessed and scored for how considerate they are in their site management and operations. The site management plans of any good construction manager.

While some people would argue that being a good neighbour is just another ploy to impress clients and win more work, it is never a bad idea to help the local community. Despite the various motivators, working in a considerate way will benefit everyone in the long run, including those who work on the sites, and so should be a key aim of scheme gives annual awards to the best performing sites, and many clients now make membership of the scheme a requirement before a project has even been won, with a certain score to aim for.

Considerate Constructor's Scheme membership is proudly displayed on many site hoardings, as shown in Figure 3.2.15.

The Scheme looks to award marks in five key areas:

● Enhancing the appearance of the industry – constructors should ensure that sites appear professional and well managed.

● Respecting the community – constructors should give the utmost consideration to their impact on their neighbours and the public.

● Protecting the environment – constructors should protect and enhance the environment.

● Securing everyone's safety – constructors should attain the highest levels of safety performance.

● Caring for the workforce – constructors should provide a supportive and caring work environment.

(CCS 2013)

These areas reflect the impact which construction sites can have on the surrounding neighbourhood, as well as looking to make sure they are as healthy, safe and environmentally friendly as possible. The criteria of the scheme set out a good approach for construction managers to try to reduce the impact of some of the problems of site management, and to try to develop good relationships with those living or working around sites.

You can find out more about the scheme on the CCS website.

Figure 3.2.15 The Considerate Constructors Scheme

3.2.5 Site management control

Like all aspects of construction management in practice, site management is not a one-off activity for good construction managers. They do not just develop the site layout before the project begins, pin it to the wall and leave it at that. Site management is an ongoing process and requires management control to ensure that the plan is being carried out and that the existing site management provisions remain the most effective and efficient for the works at any given stage.

Site management should follow the programme, and the construction manager should note or schedule, ideally on the programme itself, any events that will affect site management and require action, such as the preloading of materials or critical changes in access.

Over time, the site will develop and change as a result of many things, including the natural ongoing nature of the work:

- Trades and their stores will come on site to start work.
- Trades and their stores will leave site once their work is complete.
- Construction work will occur that affects access.
- Construction work will occur that affects loading provision.
- Unexpected things in the ground.
- Unexpectedly bad weather.
- Unexpected changes to design may be instructed.
- Unexpected changes to the programme may occur.

While most of the above issues can and should be planned for, unexpected events do happen, and the construction manager must be able to respond quickly to such changes to ensure continued smooth running of the site, by planning and implementing effective and convenient alternatives in good time.

Site inspections

Ongoing monitoring of the site should form part of a construction manager's daily activities. General site walk-rounds will quickly show an observant construction manager whether the site layout is operating efficiently or not. As a general rule a tidy site is an efficient site, and so a quick inspection will indicate whether the materials stores, access routes and waste management provision are meeting requirements.

3.14

> ## Discussion Point
>
> How could you tell whether the site welfare provision was meeting requirements?
> What about the traffic management plan?

More formally, daily and weekly **inspections** should be carried out by the construction manager in order to record any issues, which can then be addressed directly with the subcontractors involved.

A formal checklist for a weekly inspection could ask the following questions:

- Site boundary
  - Is the site boundary secure?
  - Does it give a good impression of our company?
- Traffic management
  - Are vehicles and people segregated?
  - Are barriers secure and complete?
  - Are the crossing points clearly marked?
  - Is the traffic management plan up to date?
  - Can access plant move and operate safely?
  - Are the loading points clear and tidy?
- Pedestrian management
  - Are all walkways clear?
  - Are segregating barriers secure?
  - Are walkways being used correctly?
  - Are crossing points being used correctly?
  - Are there any trailing leads across walkways?
  - Are any materials blocking access routes?
  - Is any waste blocking access?
  - Are emergency escape routes clear and signed?
- Housekeeping
  - Are all work areas clean and tidy?
  - Are there any piles of waste that should be removed?
  - Is there any significant fire risk from materials/waste?
  - Are fire extinguishers available and working?
- Materials storage
  - Are the stores clean and tidy?
  - Are all materials stacked safely?
  - Are all materials free from contamination and protected from the weather?
  - Are any materials being damaged?
  - Are materials being handled efficiently?
  - Are pallet trucks available where needed?
  - Are materials easily and safely accessible?
  - Are all containers secured?
- Plant storage
  - Are the plant stores clean and tidy?
  - Is all plant being well maintained?
  - Are drip trays in use at refuelling points?
  - Is small plant securely locked away?
- Waste management
  - Are the skips clearly labelled?
  - Is waste being placed in the correct skip?

- Is the skip area clean and tidy?
- Is there safe and easy access for skip changes when needed?

- Welfare

- Is welfare clean and tidy?
- Is there sufficient provision at breaks and lunch-times?

Formal inspections should be recorded and actioned as part of the quality management system of the site.

Where problems have been spotted – for example, if work areas are not clean and tidy – the construction manager must take action to have the areas cleared up. The removal of the scaffold from the side of the building in Figure 3.2.16 has revealed some waste materials, and these should be cleared up quickly before others join them.

The construction manager may issue a **clean-up notice** to the relevant subcontractors, who can be identified by the types of materials. Here, a mortar tub and brick guard show that the bricklayers have some tidying up to do, and the insulation quilt means that the roofers have also contributed to the mess. Depending on how serious the problem, formal clean-up notices can be issued for immediate or 24-hour action by the subcontractors.

If subcontractors do not respond in time, the main contractor can clear up the mess and charge the subcontractors for the labour used. If operatives have to pay for their mess, or stop work themselves to tidy up, they may think twice about not putting their waste in the skip next time – stopping work often means stopping making money, and charges can also add up. However, finding the labour to clear up other people's mess can be difficult for some construction managers, as

Figure 3.2.16 Waste materials that have collected under scaffolding

although general labourers can be very useful for this task they are often not costed for in the site management budget, as technically all the individual subcontractors should clear up their own mess. It can be hard to bring in labour to undertake such tasks, and so it is important that the construction manager keeps tight control of their site, to ensure good site management from the start.

This involves ensuring good **housekeeping** on the site – keeping it clean and tidy. Good housekeeping is a vital part of site management, and is legislated for within the CDM Regulations 2007 which require sites to be kept in good order, and in a reasonable state of cleanliness. A well-kept site is shown in Figure 3.2.17: materials are neatly stored and a wheelie-bin has also been provided so that operatives can easily keep the area tidy.

Facilitating good housekeeping is an important part of site health and safety management – by making it easy for people to maintain a clear work area many hazards and risks will be avoided. Mess and rubbish also attracts other mess and rubbish, so even small piles of waste should be cleared up immediately by those who made them. A clean and tidy site is more likely to stay that way.

Discussion Point

What methods could be used to ensure that good housekeeping is maintained?
What are the safety risks from poor housekeeping?

Figure 3.2.17 Good housekeeping

Most people want to work in a clean, tidy and well-managed workplace; it makes them more efficient and, when their efficiency is equal to their weekly wage, it can be a significant motivator. Consequently, a poorly managed site, badly planned and uncontrolled, can lead to demotivation and an unsafe work environment. Just one small pile of waste can grow over the course of a week into a mountain, as poor habits encourage others to follow. High standards for site management must be set early on in the project, and all operatives who come to work on the site must be made aware of these standards at their initial **site induction**, as well as penalties for not complying.

Any other issues highlighted in the inspections should be raised immediately with those concerned, and action taken where necessary. For example, if the pedestrian barriers are no longer secure, they need to be made safe as quickly as possible. Site inspections will also highlight potential health and safety and environmental issues, and so can help the construction manager take action before an accident or incident occurs.

Regular and rigorous site inspections are also a key tool in health and safety and environmental management, and are vital in helping to manage and control these parameters of production. Site management should also be an ongoing agenda item for site team and subcontractor meetings in order to prompt discussion of any forthcoming changes that could affect the current site management practices.

3.2.6 Site management: summary

Site management forms a key part of production management. Without supportive site management planning and control, the production parameters would not be met effectively or efficiently.

Site management planning is another part of production planning, and must reflect the work methods chosen for the project. Enough money must be allocated within the tender preliminary pricing to be able to meet the needs of the decisions made at these early stages, as well as ensuring that all legislative requirements will also be met. The development of the site layout needs to reflect these early plans in more detail, and to make sure that site management provisions are clearly communicated and understood by everyone working on the site.

Site management control forms a large part of the construction manager's daily activities, making sure that the plans and provisions made in the site layout are in place, and that they are still the best option as the project and site changes over time. Sites must be reassessed and re-planned as the works progress to ensure that the optimum site layout is in place, with the optimum management provisions, to ensure that optimum production can be achieved.

Discussion point comments

Discussion Point

How could pollution be caused on sites?
What may be done to mitigate against this potential environmental impact?

Poor handling and storage of pollutant materials such as lubricant and soap oils (used for cleaning shuttering), poor maintenance of plant which can leak fuel, poor concrete wash-down facilities and poor waste management practices can all cause pollution on sites.

Mitigating or reducing the potential for environmental impact means better control of pollutant materials. Locked stores should be used for oils and lubricants, and they should only be used by those also trained to manage any environmental incidents that could occur, using spill kits or similar equipment.

Plant must be maintained and a maintenance log kept to make sure that this happens on a regular basis. All refuelling should be carried out over a drip tray, again by operatives trained in environmental incident management in case a spill occurs.

Concrete wash-down should only occur in designated areas where there is no risk of the runoff getting into the surface water drainage. This should be managed across the site, and spill kits located to make sure that any incidents will be controlled.

Waste management should look to prevent materials from sinking or degrading into the ground. Hazardous waste should be contained within closed and locked skips and drums to prevent water getting in and passing into the ground.

More information on environmental management may be found in Section 3.7.

Discussion Point

How would you know if there was a right of way on the site?
What could you do about it during the project?

Public rights of way are a matter of public record and may be found on Ordnance Survey maps, via local council websites or identified through signs physically located on rights of way.

Rights of way should be identified at the very beginning of a project or during site visits, and should be allowed for in project designs. But some rights of way may not be safe during the construction work, for example, if they pass through or next to the site. Rights of way can be temporarily changed to a safer route, but any change must still be convenient for use and approved by the local council. A route may also be closed if it can be shown that it is no longer used or is supporting crime. Alternatively, scaffold fans or shelters may be constructed temporarily over the right of way, to protect users during construction work. Clear signage should be used to tell people why changes or temporary structures are in place, and to inform them of the number to call if they have any queries or concerns.

3.6 Discussion Point

What pros and cons could there be with the different boundary methods?

What production parameters should be considered?

There are three main types of boundary: timber hoarding, mesh fencing and solid panel fencing; the pros and cons of each type are outlined in Table 3.2.2.

The length of the project (its overall duration) will affect the type of hoarding selected. For a short-term project solid timber hoarding may be a waste of costs and resources. The budget available will also affect the decision, as will available materials from other sites. Environmental concerns will also be very influential, as a solid hoarding will help mitigate against noise and dust in the surrounding area.

Table 3.2.2 Comparison of site boundary types

Solid timber hoarding	Open mesh fencing	Solid metal panel fencing
Robust and solid.	Can be pushed over if not firmly fixed through the feet.	Can be pushed over if not firmly fixed through the feet.
Cannot see through unless vision panels provided – hiding trespassers and thieves.	Can see through so that trespassers and thieves can be clearly seen by passers-by.	Cannot see through – hiding trespassers and thieves.
Mitigates dust and noise transfer.	Noise and dust go straight through unless sheeted.	Mitigates dust and noise transfer.
Cannot be moved.	Flexible and can be moved as the project changes.	Flexible and can be moved as the project changes.
Can be recycled at the end of the project, but can also be used in the works.	Fully reusable.	Fully reusable.

Discussion Point

How could site access affect and potentially be a hazard to the public?
What measures could you put in place to help protect them?

Site access often means creating a new crossing point for the public on the pavement outside the site, and turning wagons and other vehicles therefore become a new hazard to the public on footpaths.

Any new access gates should be clearly signed and marked, both for the pedestrians and the drivers of any delivery vehicles. Where turning is tight, mirrors can be fitted to show drivers if anyone is walking along the pavement before they pull out. If reversing is necessary, wagons should be directed out of any gates by a qualified banksman to ensure safe manoeuvring.

If wagons are crossing pavements, the pavements may become damaged by the extra weight, and so some protection, such as road plates, may be needed to keep a level path for pedestrians.

Discussion Point

Why might traffic or pedestrian barriers be moved?
How would you manage this as the construction manager?

There are many reasons why barriers are moved: to allow deliveries to be unloaded, to help a wagon turn round, to move the MEWP from one area to another, or to walk across to another part of the site. But none of these are actually vital to the construction work. Often people have good intentions but forget to put barriers back after wagons have been unloaded or after MEWPs have been moved, and in some cases they are just taking a short cut and being lazy.

As the construction manager, a system for checking and securing barriers should form part of regular daily site inspections. An inspection in the morning before work starts will make sure that everything is set for the day, and another after lunch to check that everything is still all right. More informal approaches may also be used; asking subcontractors' supervisors to replace any barriers they see moved during the course of their day will also help keep them in the right positions. Making sure that all the operatives know why barriers are there at inductions will also help keep them in place.

As a last resort, people can also be punished for moving something put in place for the health and safety of others; it is against the law (Health and Safety at Work etc. Act 1974). The disciplinary procedures should also be included in the site induction, and consistently enforced to make sure that everyone helps in keeping barriers in the right place.

Discussion Point

What construction activities might cause a fire on site?
How do you think this could be managed?

Any construction work that involves heat, a flame or sparks has the potential to cause a fire on site. Activities may include grinding or cutting metal (for example, reinforcement or the metal studs used in partition walls) or welding. Fires can also be caused by some materials used on site which combust at certain temperatures. Smoking is also a potential fire risk, as is the more deliberate act of arson. Poor housekeeping can increase the risk if lots of flammable packaging or materials are left on site.

Fires need three things to burn: an ignition (from a welding spark), fuel (cardboard boxes from unpacking some materials) and oxygen (from the air). The three together can easily cause a fire on site.

Works that could cause a fire are often managed by a permit system – a '**hot works**' permit. Permits are issued by the construction manager to ensure that operatives are aware of the works being carried out. The permit requires the operatives who will be causing sparks to check that their work area is clear and there are no potentially combustible materials nearby, to make sure they follow their safe system of work, to have a fire extinguisher on hand, and to check the area one hour after they finish work to ensure that a fire has not been smouldering.

Discussion Point

How would you show the emergency escape routes on site?
How would you make sure they were kept clear?

Emergency escape routes should be signed as they are in a completed building, with standard emergency signage that will glow in the dark and emergency lighting switched on should there also be a main power failure. They should also be identified on the site layout drawing if possible, or, if that would make the layout too busy and hard to read, a separate emergency escape plan can be created and posted on noticeboards in welfare cabins and on site.

Management of emergency routes should form part of regular daily and weekly inspections, and be informally checked every time construction managers tour sites, and their management should be highlighted as a concern for all operatives on site at the induction. Problems of waste or of materials being stored in or blocking the emergency routes should be raised with the relevant subcontractors as soon as they are identified, with immediate clean-up notices issued if needed.

Discussion Point

Skips need access for people, machinery and vehicles.
How would you manage this area to keep it safe?

Appointing a general operative or trade foreman to a skip-management role will mean that someone on site all the time is keeping an eye on the skips to make sure the area is kept in good order. The skips should be inspected as part of the regular daily and weekly inspections, as well as informally whenever they are passed.

Skip wagons should be banked (manoeuvred with a banksman to guide their way) whenever they are to be delivered or collected, and an operative should be on hand to move any segregation barriers to allow access to the skips, putting them back carefully once the skip has been changed.

Discussion Point

What benefits could there be to developing good relationships with the neighbours?

The local neighbourhood can also be a source of useful knowledge about the site, as well as several pairs of eyes that can watch the site when the workforce has gone home for the day. Developing good relationships within the local area can also help develop a good local reputation, which may help win work in the future.

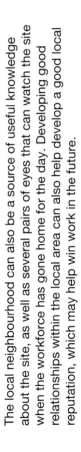

Discussion Point

Do you think construction companies should invest in the local communities they work in? Why?

Investment in the local community will help develop a company's reputation as a key member of that community. This may help them to win work in the future, as clients become more concerned about their own reputations, as well as what they can gain in local goodwill while working on the construction project.

It should also be said that anything companies can do to help local areas improve and provide opportunities for people through local apprenticeships or training is a good thing, and should be encouraged no matter what the underlying reasons may be.

Discussion Point

How could you tell if the site welfare provision was meeting requirements?
What about the traffic management plan?

If site welfare is clean and tidy at the end of the day then it is meeting requirements. If it is dirty and overflowing there may not be enough provision for the number of workers currently on site. The best way to find out is to have a look around the site and talk to people to see if they have enough places to wash or enough seats available at breaks and lunch-times.

If the traffic management plan is not working there are likely to be traffic jams and arguments on site on a daily basis. Barriers will be frequently moved to allow access for wagons, and store areas may become clogged or spring up in unauthorised areas. Again, the best thing to do is talk to people, especially those taking deliveries on a regular basis to see that the system is working properly. Checking outside the site will also help to make sure that wagons are not waiting and blocking surrounding roads, or creating hazards for pedestrians when turning into the gates.

3.15

Discussion Point

What methods could be used to ensure that good housekeeping is maintained?
What are the safety risks from poor housekeeping?

Good housekeeping can be maintained by the following:

- Diligent management ensuring that people are tidying up as they go.
- Providing bins for easy clean-up.
- If they are subcontractors, writing it into their contract, and making it possible to reclaim money if others have to tidy up their mess.
- Educating the workforce through tool-box talks and site inductions.
- Making good housekeeping a site rule.
- Punishing those with untidy work areas and making them stop work to tidy up before they can carry on.

The key risks from poor housekeeping are slips, trips and falls. Many site accidents are caused in this way and can result in twisted ankles, sprains or even breaks. There may also be a risk of fire if flammable materials are piled up on site.

3.3 Time management

3.3.1 Introduction

Time is managed through the use of the construction **programme**. Planning is the management process that develops the plan for the project; the programme is the communication of this plan.

As we have seen, different programmes will be produced at different stages of the project:

- Pre-tender
- Pre-contract
- Operational.

We are going to concentrate here on operational planning and how construction managers can follow the planning process to create the programmes they need to help them manage and control the production parameter of time on sites.

Operational planning also looks at different levels of detail for the work and produces the following:

- Project programmes
- Three-month look-ahead programmes
- Weekly programmes.

Remember: the method for communicating the *plan* also needs to make *control* as easy as possible for construction managers as the works progress.

Another key issue to note is that there is not always just one right answer in planning and programming. Two industry planners may come up with quite different programmes for a project, each applying their own unique knowledge and experience. The overall aim is to meet the production parameters for the project within the practical constraints of construction activity, and there may be a number of ways to do this in practice.

This section will first look at how construction programmes are produced: the planning aspect of time management. This will first be examined through the production of simple bar or Gantt charts, which will be developed to include logic. Logic enables the identification of **critical path** and **float**, both of which are vital tools for construction managers when planning and controlling the work on site. The section will then look at how control happens in practice and how programmes may be used as a management tool.

3.3.2 Producing a construction programme

Construction time management is often carried out using computer software to produce a bar or Gantt chart. Often when people talk about 'the programme' it is the Gantt chart they are referring to.

A basic Gantt chart programme is shown in Figure 3.3.1. This provides a quick representation of what Gantt charts are, and includes milestones, activities, durations and sequence.

		Date														
Activity	**Duration**	1	2	3	4	5	6	7	8	9	10	11	12	13	14	15
Start	0 days															
Activity A	2 days															
Activity B	4 days															
Activity C	2 days															
Activity D	1 days															
Activity E	4 days															
Activity F	3 days															
Completion	0 days															

Figure 3.3.1 A basic Gantt chart programme

Most construction programming is now carried out using computers, and there are a number of different computer software planning packages available. For example, MS Project, found within the Microsoft Office Suite, may be used to create simple programmes, while more complex project management software packages, such as ASTA Powerproject, have more detailed planning functions and may be used to produce more detailed programmes.

As with any computer software, experimenting with it will help you gain skills, and any of the exercises in this section will give you a starting point to try out the software available to you at your college or university. Most are intuitive programmes that prompt you where possible to make sure your programme is robust, and they often have useful help functions if you get stuck.

Programmes used to be produced once the planning process had been undertaken, the construction methods selected and all parameters considered in detail. The flexibility of computer software means that the programme can be produced as part of the production planning process, and various approaches and construction methods can be tested in terms of time management, and within the other production parameters, using the software itself.

The following steps are taken in the production of a construction programme:

- Establish key dates and **milestones.**
- Establish key **activities.**
- Establish durations.
- Allocate resources.
- Establish the sequence.
- Apply **logic.**
- Analyse for critical path and float (logic).

3.3.3 Establishing key dates and milestones

Operational programmes are developed from the pre-tender and pre-contract plans, and must reflect the same key elements. This is known as being **back-to-back** and ensures that the production parameters taken into consideration before the work was won are maintained, to ensure that overall plans are achievable.

This means that key dates such as contractual start and completion dates must be included in the operational programme. There may also be sectional handover dates, where a certain part of the project must be completed and ownership handed over to clients before the final completion date is reached.

Important dates such as this are known as project milestones. Planning may be carried out backwards from such milestones to ensure that the project meets these critical dates. If such dates are missed there can be serious repercussions for contractors.

Milestones do not have a time duration and are assigned zero days; they are represented by a diamond, as shown on the start and completion activities in Figure 3.3.1.

Discussion Point

What other key dates or time periods may need to be included? Think about key dates for specific types of projects, for example, schools.

3.3.4 Establishing key activities

The next step is to determine the key activities or work tasks to be carried out. Activities are the tasks or construction work to be undertaken which have a time and resource value associated with them.

Activities may be listed in various levels of detail. The master programme for the operational phase of the project will contain activities listed at a relatively limited level of detail – one activity title may cover a lot of different work tasks in reality. More detailed programmes, known as 'short-term' or 'look-ahead' programmes, can cover different time-scales. For example, a programme may plan works for the next three months or plan only the next week's work, in which even more detail of the planned construction activities may be included. All of these programmes must still be back-to-back to make sure key milestones are met.

Here it is important to remember the *function* of the programme and how it will be controlled. The level of detail in the programme should reflect the level of detail needed to effectively plan and control the time management of the works on site. If the programme is to be used to plan daily work then the activities should identify the activities at the level of the different resources used to carry them out.

For example, if we consider the first steps in constructing a house, the three-month look-ahead programme may simply list the following elements:

- Drainage
- Foundations
- GF slab
- Brick block to first-floor level.

But the weekly operational programme may show the following elements:

- setting out;
- excavate, lay and backfill drainage;
- excavate and blind foundations;
- pour concrete foundations;
- substructure blockwork and DPC;
- stone up for slab;
- DPM/gas membrane/insulation;
- reinforcement;
- pour concrete slab and cure;
- brick/block first lift;
- scaffolding;
- brick/block second lift.

All of these activities add up to the same thing in construction terms, but the difference is in the information needed for the detailed management control of the work. The weekly operational programme provides a 'to-do' list for the construction manager which is easily broken down into different resources: labour (groundworkers and bricklayers), and plant (the excavator and scaffold), and materials (concrete, DPC/DPM, insulation). This makes the programme a more useful construction management tool in terms of planning and control.

Activities are listed on the programme down the left-hand side, as is shown in Figure 3.3.2.

Exercise F

What activities need to be added to the above list to complete the house?

Assuming a two-storey detached construction, list the activities to help the future development of weekly programmes.

Suggested solutions to the exercises may be found in the Appendix.

Activity	Duration	Date															
		1	2	3	4	5	6	7	8	9	10	11	12	13	14	15	16
Setting out																	
Excavate, lay and backfill drainage																	
Excavate and blind foundations																	
Pour concrete foundations																	
Substructure blockwork and DPC																	
Stone up for slab																	
DPM/gas membrane/ insulation																	
Reinforcement																	
Pour concrete slab																	
Brick/block first lift																	
Scaffolding																	
Brick/block second lift																	

Figure 3.3.2 Adding activities to the programme

3.3.5 Establishing durations

Durations are the *length of time* an activity will take with a specific allocation of resources. For example, it may take a decorator one week to paint the interior of one house, so it should take two decorators half a week.

Once the activities have been determined to the required level of detail, time durations must be applied against them depending on the resources allocated.

In order to find out how long an activity will take, the following formula is used:

$$\frac{Quantity}{Output\ per\ hour} = Hours\ (the\ duration)$$

Quantity = the quantity of the item, for example, the m² of walls to paint inside the house.
Output per hour = the m² a decorator can paint in one hour.
Hours = the total amount of time it will take to paint the house.

Based on this information, the number of days the activity will take can now be calculated by inputting the answer into the following formula:

$$\frac{Total\ hours}{Number\ of\ hours\ per\ day} = Total\ days$$

The number of days the activity will take may now be included in the programme, depending on the average number of hours worked per day. Some sites adhere to the National Working Rule Agreement of 39 hours per week, but some operate longer working days. In addition, be aware that many sites close early on Fridays and so this needs to be taken into account when planning the working week.

Outputs per hour may be obtained from many different sources. Large organisations often have their own planning departments which will gather performance data feedback from the projects they work on over time. This often provides an accurate set of outputs that may be applied to different activities in different circumstances.

Data may also be obtained from *Spon's Architects' and Builders' Price Book* which contains durations for the different activities alongside the current year's pricing. Experience can also be valuable in assigning durations to activities, as well as requesting data from the various subcontractors who will actually be involved in the work.

As long as a reliable source may be found for the output per hour, durations for any activities can be calculated.

Worked example

Calculate the time it would take a joiner to hang 26 single-leaf doors.

It will take one joiner 45 minutes to hang one door.

Step 1: Turn the minutes into hours. To do this, divide by 60, as there are 60 minutes in one hour:

$$\frac{45\text{mins}}{60} = 0.75\text{hrs}$$

It is better to use decimal hours than minutes, as this corresponds to price books and makes sure everyone is using the same way of working.

Thus one door takes 0.75 hours to hang.

But this is hours per output, and for the standard formula we need to know output per hour.

So rearrange the equation:

$$\frac{1}{0.75} = 1.34\text{doors/hr}$$

Step 2: This can then be included in the formula:

$$\frac{\text{Quantity}}{\text{Output per hour}} = Hours$$

Thus:

$$\frac{26\text{doors}}{1.34\text{doors}/\text{hr}} = 19.40\text{hrs}$$

Step 3: From the calculation, it will take 19.40 hours to hang the 26 doors, so add this into the following formula:

$$\frac{\text{Total hours}}{\text{Number of hours per day}} = Total\ days$$

Activity duration: 19.40hrs
Typical working day: say 8.0hrs (39-hr weeks)

$$\frac{19.40\text{hrs}}{8\text{hrs per day}} = 2.43\ days$$

Time is usually only recorded to the full day, so we would enter the above activity as taking three consecutive days.

Establishing durations is not an exact science, so don't be fooled by the above equations. Planning construction work also needs to be 'slacker' than estimating costs for the same work – it needs to allow for problems and glitches to arise, such as operatives or deliveries not

Activity	Duration	Date															
		1	2	3	4	5	6	7	8	9	10	11	12	13	14	15	16
Setting out	1 day	▮															
Excavate, lay and backfill drainage	1 day		▮														
Excavate and blind foundations	1 day			▮													
Pour concrete foundations	1 day				▮												
Substructure blockwork and DPC	1 day					▮											
Stone up for slab	1 day						▮										
DPM/gas membrane/insulation	1 day							▮									
Reinforcement	1 day								▮								
Pour concrete slab	1 day									▮							
Brick/block first lift	3 day											▮	▮				
Scaffolding	1 day													▮			
Brick/block second lift	3 day															▮	▮

Figure 3.3.3 Adding durations to the programme

turning up on time. In many cases there is not enough information to calculate durations precisely, so estimates are often required based on a combination of factual data, advice and experience.

Durations are added to the programme along the time or x axis. They are shown as bars for the durations of each activity, as illustrated in Figure 3.3.3.

Duration exercises

Calculate the durations for the following activities in Exercises G to J.

For all tasks, assume a working week of 8 a.m. to 5 p.m. Monday to Thursday, and 8 a.m. to 3 p.m. Friday, one hour for lunch and breaks each day.

Exercise G

An excavator and driver have a performance output of 25m³ per hour (constant); 10,000 cubic metres of spoil need to be excavated to prepare the sites. How long will this take?

Exercise H

A wall is required to the front of a domestic property garden. The specification is as follows:

- 1.8m high.
- Length of the wall will be 25m.
- Coping to the full length.

Assume there are 59 bricks per metre squared and the bricklayer can lay 50 bricks per hour.

The coping stones are 600mm long and it takes the bricklayer 10 minutes to lay one coping stone.

How long will it take one bricklayer to complete the wall?

It takes a floor layer 0.46hrs to lay 1m² of flooring. How long will it take one floor layer to complete the first floor which is a total of 425m²?

Suggested solutions to the exercises may be found in the Appendix.

3.3.6 Allocating resources

For construction managers, allocation of resources is concerned with optimisation of production. As demonstrated in the calculation of durations, the number of resources added to an activity affects its overall duration and therefore the levels of production on the project.

However, while adding more resources can speed up work, it is not always so straightforward. In many cases it is not safe for too many people to be working on one activity or in one area at the same time, and quality of work can also suffer. While textbooks and price books indicate productivity constants, and therefore the time taken for one man or one machine to complete a task, it is only by drawing on experience that judgements can be made about optimum resource allocation. How many 2+1 gangs of bricklayers to build a large four-bed detached house? How many machines to complete the substructure and external works? How many 2+1 gangs of bricklayers to employ on a speculative housing project of 100 units? How many machines ditto?

Often over-allocation of resources is not economic. In addition, the motivation of operatives can be affected by resource allocation, since they like to see a period of continuity of work in front of them, on which they can earn good money uninterrupted by other people. If there are too many people in one area 'fighting' for work, tradespeople may just walk off the job and find work elsewhere.

Alternatively, if too few resources are allocated, the work will progress too slowly and not meet the key dates of the programme.

Resource allocation must be considered early in the pre-tender planning stage, to meet the production parameters specified by the client, including time and cost, and to support the chosen construction methods that have been used to create the tender price. Often, the construction manager will receive information about resource allocations from estimating teams when they start operational planning. Rather than significantly reassessing resource allocations, the construction programme should allocate resources to meet or better the costs of those previously planned to ensure that there is no overspend.

For construction managers, resources are as follows:

- Labour
- Plant
- Materials.

Resources of all three types can be allocated to each activity within the programme, either in the form of a resource histogram or percentage allocation of available resources to tasks, depending on the programming software being used. An example of how they may be added easily is shown in Figure 3.3.4. Here the plant and labour has been added to each activity in shorthand; for example, 'exc' refers to the excavator and g/wk are groundworkers. The labour levels are shown on the histogram below the programme, the different colours indicating different trades.

From this information, the construction manager can easily see that while two groundworkers are needed for the majority of the work, the bricklayers are needed less consistently. They have to visit the site early for the substructure blockwork, then return for the superstructure brickwork and have one day off site while the scaffold is constructed. If there were more plots on the site, the construction manager could use the information from resourced programmes to produce a sequence across all plots that provides **continuity of work** for both labour and plant, making their site as efficient as possible. For both labour and plant, continuity of work is vital: it is ineffective to complete an activity early if the labour and plant are then standing idle, waiting for the next activity to become available. Other activities may be used for labour and plant to drop on to if they do become free – these activities are known as **hospital jobs**.

Such detail is often only identifiable at the operational level, and so it becomes the role of the construction manager to allocate resources to their weekly operational programmes to ensure the highest levels of production efficiency.

Resources form a significant part of project cost, and their mismanagement can quickly use up profits. Resources must therefore

Activity	Duration	1	2	3	4	5	6	7	8	9	10	11	12	13	14	15	16
											Date						
Setting out (1 eng + 1 g/wk)	1 day																
Excavate, lay and backfill drainage (exc + 2 g/wk)	1 day																
Excavate and blind foundations (exc + 2 g/wk)	1 day																
Pour concrete fdns and cure (conc + 2 g/wk)	1 day																
Substructure blockwork and DPC (2+1 bricklayers)	1 day																
Stone up for slab (Exc + roller + 2 g/wk)	1 day																
DPM/gas membrane/insulation (2 g/wk)	1 day																
Reinforcement (2 g/wk)	1 day																
Pour concrete slab and cure (conc + 2 g/wk)	1 day																
Brick/block first lift (2+1 bricklayers)	3 day																
Scaffolding (2 scaffolders)	1 day																
Brick/block second lift (2+1 bricklayers)	3 day																
	3																
Labour Resource Histogram	2																
	1																

Figure 3.3.4 Adding resources to the programme

be clearly shown on the programme to ensure that they are allocated as originally planned.

3.3.7 Establishing the sequence

Sequence refers to the order in which the activities should be carried out. This can be influenced by several different considerations, including the key production parameters.

Sequence is, however, always driven by technical detailing of the construction work. For example, the design details of the ground-floor slab will dictate the sequence of stone/insulation/DPM/concrete installation.

Consideration of the practical nature of construction work is also necessary, and the construction manager often needs to draw on their experience to decide how to plan their work. For example, some work should be left until as late as possible within the project, so that it is not damaged by following trades. Floor screed is often vulnerable to damage and so should be pushed as late in the programme as possible, but screed also needs to be laid early enough to lose sufficient moisture before floor finishes such as vinyl are laid on top. Therefore there will be a balance between the two to optimally sequence the screed within the construction programme.

In some cases the earlier the installation the better. Achieving live services (e.g. electricity and mains water) early on in the programme can help with the testing and commissioning of service provisions, and also bring heat and light to the project earlier to help with drying out and improve working conditions for later trades.

Both technical and practical knowledge is necessary for the construction manager to ensure that activities are planned in the correct and optimum sequence, and will not result in re-work or work of poor quality.

Sequence exercise K

Consider Figure 3.3.5: this shows an internal disabled toilet fit-out. Think about the space as a 'shell' – the blockwork walls, concrete soffit ceiling and concrete floor slab. Out of picture are a hand-drier, suspended ceiling and emergency alarm.

Carry out the planning for the fit-out of the shell to the finish shown in Figure 3.3.5, at the level of detail needed to produce weekly construction programmes for the work:

- List the activities needed to fit out the toilet.
- Place these activities in the optimum sequence for the work to meet the technical requirements and minimise damage to existing work.
- List the trades that would carry out each activity, and decide if everything can be done by each trade in only one visit.

Suggested solutions to the exercises may be found in the Appendix.

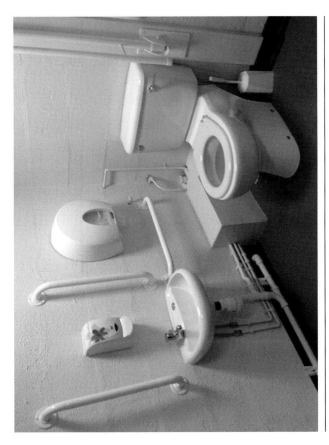

Figure 3.3.5 Internal toilet fit-out

3.3.8 Applying logic

Logic is the term for links made between activities to keep the sequence of works in order and which represents the relationships between them. Logic makes programmes far more robust and enables them to be used as time management tools which may be used for planning and control, rather than just representations of the original plan.

Within the Gantt chart, logic links are shown in the form of *arrows* linking the activities. These arrows represent the relationships between the different activities. Care must be taken to make sure *all* the activities are linked together from project start through to project completion to create a programme that is a functional management tool, rather than just a one-off plan.

For example, the following relationships may be shown on a programme through logic; the numbers in brackets represent the duration of each activity:

- The programme starts with Activity A(2).
- Activities B(6) and C(2) cannot start until Activity A has been completed.
- Activity D(4) cannot start until Activity B has been completed.
- Activity E(2) cannot start until Activity C has been completed.
- Activity F(1) cannot start until both Activities D and E have been completed.

This programme would be as shown in Figure 3.3.6.

Activity	Duration	Date														
		1	2	3	4	5	6	7	8	9	10	11	12	13	14	15
Start	0 days															
Activity A	2 days															
Activity B	6 days															
Activity C	2 days															
Activity D	4 day															
Activity E	2 days															
Activity F	1 day															
Completion	0 days															

Figure 3.3.6 Adding logic to a programme

Exercise L

On the programme paper template provided in the Appendix, produce the logic-linked programme for the relationships listed below. You can also produce the programme using computer planning software – links can often be made by dragging the mouse between the two activities to be linked; be careful to ensure that the link travels in the correct direction, always moving the programme forward. The software help packages and tutorials will also be able to show you the specific instructions for the software you are using.

Again, the duration of each activity is shown in brackets:

- Activities B(6) and C(3) cannot start until Activity A(2) has been completed.
- Activity D(4) cannot start until Activities B and C have been completed.
- Activity E(3) cannot start until Activity B has been completed.
- Activity F(1) cannot start until E and D have been completed.

Suggested solutions to the exercises may be found in the Appendix.

Different logic relationships

In the above programmes, all the relationships have been end to end; i.e. the previous activity has been completely finished before the next activity in the sequence can start.

Within construction operations, relationships are not always that simple. Quite often the next activity can commence before the previous activity has been fully completed, and some activities can occur at the same time.

Construction activities may be as follows:

- End to end – one after the other.
- Concurrent – both happening at exactly the same time.
- Overlapping – part of each activity is happening at the same time.

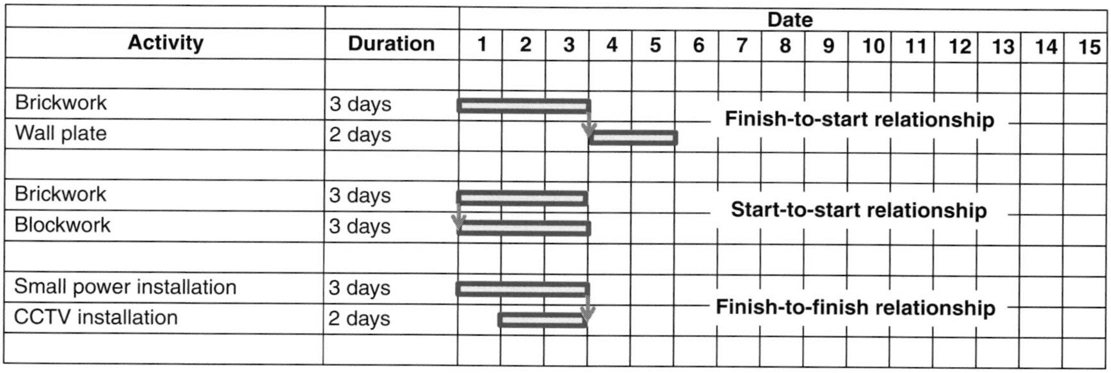

Activity	Duration	Date														
		1	2	3	4	5	6	7	8	9	10	11	12	13	14	15
Brickwork	3 days															
Wall plate	2 days															
Brickwork	3 days															
Blockwork	3 days															
Small power installation	3 days															
CCTV installation	2 days															

Finish-to-start relationship

Start-to-start relationship

Finish-to-finish relationship

Figure 3.3.7 End-to-end and concurrent activities on the Gantt chart

These different relationships are shown on Gantt charts using logic, which both defines and maintains these different relationships within the programme.

Complicated relationships are very common within construction activities, and so using the correct logic within programmes results in a more realistic and therefore useful time management tool.

End-to-end activities

End-to-end activities mean that the preceding activity must be 100 per cent complete before the succeeding activity can start. They use a **finish-to-start** logic link.

For example, the completion of a blockwork wall before the wall plate can be fitted.

End-to-end activities are illustrated on the Gantt chart in Figure 3.3.7.

Concurrent activities

Concurrent activities occur at the same time and may use either a **start-to-start** logic link or a **finish-to-finish** logic link, depending on the relationship between the activities.

Some concurrent activities start at the same time. For example, the two skins in a cavity wall could be broken down into two different activities as there are different material resources involved, but in practice they start and are constructed together.

Some concurrent activities have to finish at the same time. For example, the components of an electrical installation will need to complete at the same time to allow for power-on and testing to be carried out to the system as a whole.

Concurrent activities are illustrated on the Gantt chart in Figure 3.3.7.

Overlapping activities

The logic relationships between overlapping activities are more complicated.

For example, if a large raft foundation is being constructed then the excavator can be working in one section, the shuttering following behind and the reinforcement following behind that, as long as safe working space is maintained.

To demonstrate concurrent activities, *start-to-start* or *finish-to-finish* logic links may be used, but these links will also incorporate a time allocation – referred to as either **lead** or **lag**, depending on the relationship shown.

Lead is the amount of time the previous activity must have completed *before* the current activity can start.

For example, the shuttering cannot follow immediately after the excavation, as the machine will require some time to move ahead and leave a safe working space for the joiners. This link will therefore require a lead, say, two days, on the start-to-start link.

Lag is the amount of time *after* completion of the previous activity that the current activity needs to be completed itself.

For example, the shuttering will not be able to be completed at the same time as the excavation, as the machine will need to complete and move out of the way first. This can be shown using a lag, say, two days, on the finish-to finish-link.

Allocating lead and lag durations is not a precise calculation, and should be based on practical and technical knowledge of the activities concerned. They should enable the completion of work in an efficient, effective and safe manner, as well as making best use of the available resources.

Overlapping activities are illustrated on the Gantt chart in Figure 3.3.8.

Lead and lag may also be applied to **finish-to-start** logic links.

Rather than linking the start of two activities, this relationship links the finish of one activity to the start of the next. Therefore, the lead becomes a negative lag. Rather than the amount of time the first activity has completed before the next can start, the relationship now depends on the amount of time the first activity has left to complete, before the next can start.

An example may be seen in Figure 3.3.9.

Some computer programs (including MS Project) only allow positive and negative lag to be added to the logic links, rather than any lead, so more complex relationships involving both lead and lag cannot be

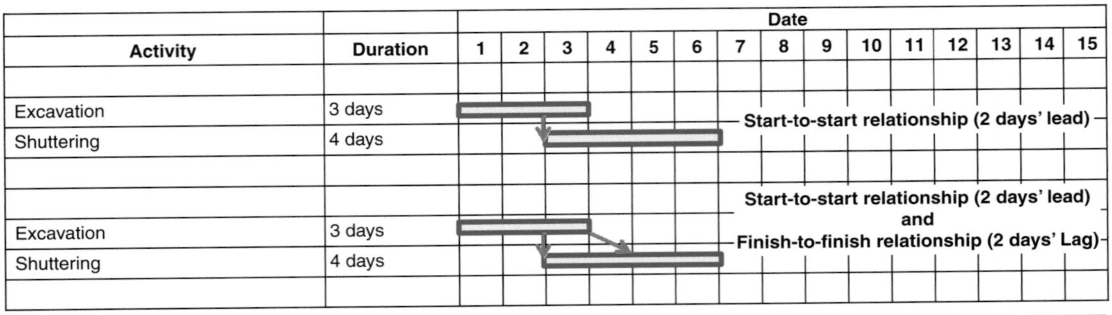

Activity	Duration	Date
		1	2	3	4	5	6	7	8	9	10	11	12	13	14	15

Activity	Duration
Excavation	3 days
Shuttering	4 days
Excavation	3 days
Shuttering	4 days

Start-to-start relationship (2 days' lead)

Start-to-start relationship (2 days' lead)

and

Finish-to-finish relationship (2 days' Lag)

Figure 3.3.8 Overlapping activities on the Gantt chart

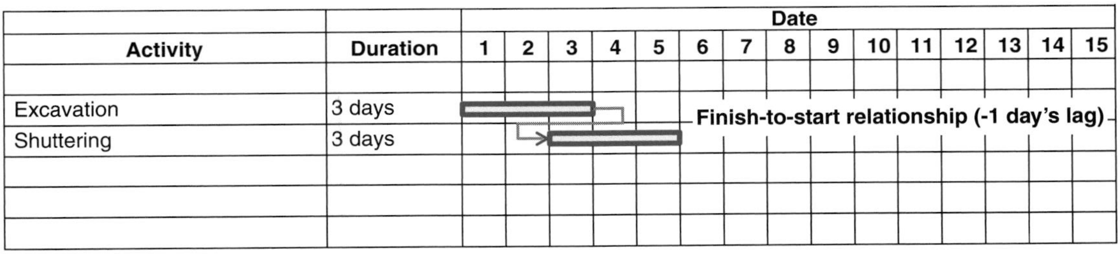

Activity	Duration	Date														
		1	2	3	4	5	6	7	8	9	10	11	12	13	14	15
Excavation	3 days															
Shuttering	3 days															

Finish-to-start relationship (-1 day's lag)

Figure 3.3.9 Use of negative lag instead of lead

added to activities. Instead, negative lag may be used to illustrate the amount of work that has to be completed before the next activity can commence in place of lead.

When constructing internal partition walls within large projects, the electrical works that provide sockets and light switches on these walls need also to be installed at the same time. Therefore internal walls involve two trades in three stages:

- Fix studwork and first side plasterboard walls (six days).
- Fit conduit and back boxes for electrical installation (nine days).
- Fix second side plasterboard walls with cut-outs for sockets (three days).

It is possible on large projects for these three activities to overlap given the different resources needed, and so lead and lag have been assigned as follows, relative to the overall internal wall duration:

- One-third of each activity is to be completed before the following task begins (lead).
- One-third of each activity is to be completed after the previous task has been completed (lag).

Produce a Gantt chart programme for the above works, showing the logic relationships and the lead and lag as noted.

Suggested solutions to the exercises may be found in the Appendix.

Critical path and float

Logic also enables the management of two key elements of time management: critical path and float. Both are identifiable within a fully logic-linked programme with a set start and completion date.

Float may be defined as any *spare time* available for an activity.

The critical path is the route along which all activities must be started as soon as previous activities are completed, and need all the available time until the next event. The critical path runs through all activities with no float, and passes through all events which have the same earliest and latest times – these are called **critical activities**.

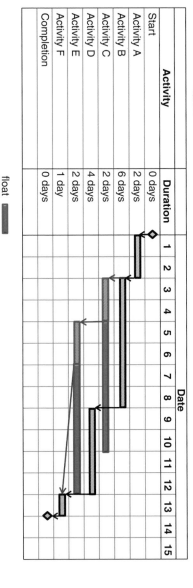

Activity	Duration	Date														
		1	2	3	4	5	6	7	8	9	10	11	12	13	14	15
Start	0 days															
Activity A	2 days															
Activity B	6 days															
Activity C	2 days															
Activity D	4 days															
Activity E	2 days															
Activity F	1 day															
Completion	0 days															

float ▭

critical path →

Figure 3.3.10 Identifying critical path and float

There may be more than one critical path in a programme – but there will always be at least one.

Many computer software packages will automatically identify the critical path and float within a programme at the click of a button, but they may also be identified from analysis of the programme.

For example, the programme produced earlier for Figure 3.3.6 has now been updated to show float and critical path in Figure 3.3.10.

In Figure 3.3.10, two activities (C and E) have been identified as having float, shown as smaller bars extending behind the activity bars. The other activity bars and their logic are now shown in red, indicating that they form the critical path.

Let us look first at the critical path. This can be traced back from the completion milestone and, following the above definition, includes any activity that does not have any 'slack' in its logic. Each activity now outlined in red has a start-to-finish relationship with the activity or activities that precede it, and they need all the available time to complete and still maintain the position of the completion date.

If any one of these activities is delayed or changes duration, it will have an impact upon the completion date, as shown in Figure 3.3.11.

Activities C and E, although they too have finish-to-start relationships with their adjacent activities, also both have float. Again, tracing back from the completion milestone, Activity E may be seen to have six days' float – spare time in which that activity can be completed and not affect the completion date. In practical terms this means that Activity E can actually start four days later (6 days – 2 days' duration = 4 days) than originally planned (on day 11), as is shown in Figure 3.3.12.

Tracing the logical relationship back up the sequence, Activity C may also be seen to have six days' float – this is the spare time in which C can be completed, but also then allow E to be completed so as not to affect the final completion date of the programme. The relationship between C and E must also be included in the float calculations. In practical terms, Activity C can actually start four days later (6 days – 2 days' duration = 4 days) than originally planned (on day 9) as shown in Figure 3.3.13,

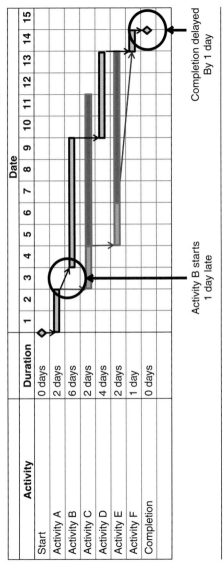

Activity	Duration
Start	0 days
Activity A	2 days
Activity B	6 days
Activity C	2 days
Activity D	4 days
Activity E	2 days
Activity F	1 day
Completion	0 days

Completion delayed
By 1 day

Activity B starts
1 day late

Figure 3.3.11 Delay in the critical path

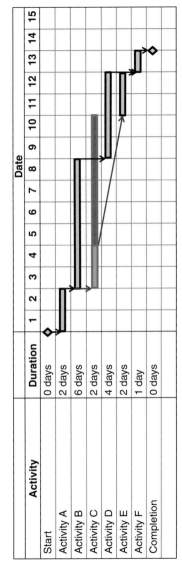

Activity	Duration
Start	0 days
Activity A	2 days
Activity B	6 days
Activity C	2 days
Activity D	4 days
Activity E	2 days
Activity F	1 day
Completion	0 days

Figure 3.3.12 Activity E delayed start

Activity	Duration
Start	0 days
Activity A	2 days
Activity B	6 days
Activity C	2 days
Activity D	4 days
Activity E	2 days
Activity F	1 day
Completion	0 days

Figure 3.3.13 Activity C delayed start

which will then still enable Activities C and E to be completed without affecting the completion date.

If this happens in practice, as shown in Figure 3.3.13, *all* activities are now on the critical path and must be completed within the allocated time for the work to complete on programme.

Activity	Duration	Date														
		1	2	3	4	5	6	7	8	9	10	11	12	13	14	15
Excavation	4 days															
Shuttering	6 days															
Reinforcement	3 days															
Concrete	1 day															

Figure 3.3.14 Float and critical path in overlapping activities

Logic exercise N

Analyse the Gantt chart you produced for Exercise L for float and critical path. Which activities are on the critical path? Which activities have float?

Suggested solutions to the exercises may be found in the Appendix.

More complicated programmes, including activities with logic relationships involving lead and lag, can also be analysed for critical path and float. In these cases, the critical path can run through just part of an activity, as is shown in Figure 3.3.14.

In Figure 3.3.14, it will be seen that the excavation works are critical for the first two days, but then the remainder of the activity has some float, as it does not need to complete until day 6 to allow for the shuttering work to complete after it. The shuttering work is critical, but only the last day of the reinforcement work is also on the critical path. There is a day's float at the start of the reinforcement work, where work could start earlier and then free up a subsequent day, should the labour be needed on other tasks elsewhere on the site.

This is the kind of information that is of most practical use to construction managers, and can help them put the plans into action on site. For example, when allocating resources at the operational level it is helpful to know that there are a few days spare in the programme to move labour or plant on to other tasks for a short time, yet still achieve the completion date.

Using logic correctly shows what activities are critical and where there is room for manoeuvre. Critical activities must be closely managed to ensure that they meet their programmed dates – any delay to them may cause a delay to the project as a whole. Activities with float have some breathing space and are not as urgent, but if they are left too long they can use up the entire float and become critical themselves.

Logic exercise P

Analyse the Gantt chart you produced for Exercise M for float and critical path. Which activities are on the critical path? Which activities have float?

Suggested solutions to the exercises may be found in the Appendix.

3.3.9 Controlling time: monitoring the programme

The main reason for producing a programme which is fully logic linked is to use it for monitoring and control once the work has begun, making the programme a useful production management tool rather than just a one-off plan.

Control of time involves checking the planned works on the programme against the actual works that have been completed on site. In order for this to be effective, the construction manager must regularly monitor works as they progress and record on a regular basis, usually weekly, how much has been completed.

This may involve the following:

- Keeping a daily **site diary** to record the number of operatives on site and the areas in which they were working. This diary should also record any key events that may have delayed work, such as bad weather.
- Marking up drawings to show work completed.
- Taking progress photographs.
- Producing a weekly progress report.
- Marking up the programme to show current **programme position**.

Progress is usually reported as a percentage completion of the activity. Estimations of progress made on site are necessary where an activity has not been fully completed. For example, if there has been an issue with the piling rig and only 25 of the 100 piles planned for the week have been completed, then this activity is only 25 per cent complete.

But in some cases this can be rather trickier; for instance, a programme activity might be 'first-fix electrical', but without detailed knowledge of what this would include (e.g. chasing, conduit, containment, cable trays) it can be hard to produce an accurate estimate. But a good understanding of the work to be done can be found in the drawings and specifications, and a *realistic* estimate can always be made.

However, be wary of reassurances that a work activity is totally complete until it has been verified; very often little bits are left undone which can all add up to problems at the end of the project. Any monitoring must be accurate and the control of the programme must be a *true* reflection of the project out on site.

Programmes may be marked up by hand on a weekly basis, drawing a progress line down the Gantt chart and reviewing the activities behind or ahead of programme. However, it can be difficult to see from a hand-marked-up programme what effect (if any) the actual works may have had upon the critical path or completion date if they have not been completed as planned.

Computer software-produced Gantt charts can be updated to show actual completed works against those planned, and are also able to do much more. At the click of a button the actual works can be recorded,

Completed activities are shaded

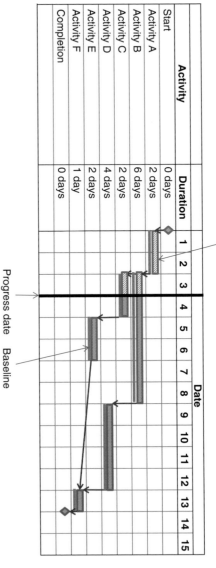

Activity	Duration	Date
		1	2	3	4	5	6	7	8	9	10	11	12	13	14	15
Start	0 days															
Activity A	2 days															
Activity B	6 days															
Activity C	2 days															
Activity D	4 days															
Activity E	2 days															
Activity F	1 day															
Completion	0 days															

Progress date   Baseline

Figure 3.3.15 Progressing the programme

and any remaining works rescheduled through the logic sequence to show revised start and finish times, and possibly even a change to the overall completion date.

For Gantt charts to be used in this way, they must be linked by *fully* logic linked, and *all* activities must be linked together through the correct relationships, from commencement to project completion. If the programme is not fully linked, when it is rescheduled the activities will not remain in sequence, and the original planning can be lost, making unnecessary re-planning work for the construction manager.

An updated or 'progressed' programme may be seen in Figure 3.3.15. Several items have been added to the programme to enable it to be progressed and rescheduled.

First, the programme has been **baselined** – the very first plan is captured and stored – before progress can be entered. This may be seen in Figure 3.3.15 by the grey lines under the main activity bars. Using computer software, the baseline can easily be captured at the click of a button.

Progress may then be added to each activity using the software. In the example above, Activity A is 100 per cent complete, Activity B is 17 per cent complete and Activity C is 50 per cent complete.

Once the logic-linked programme has been progressed it can be rescheduled, again using the computer. This automatically makes adjustments to the programme, depending on the work that has been completed to date against the plan.

In Figure 3.3.15, the actual works completed on site still match the programme baseline, and so the project is still *on-programme* to complete within the time-frame as planned.

But if the planned activities have not been completed, this rescheduling may produce a new, later, completion date. In Figure 3.3.16, insufficient work has been carried out (none of Activity B has been

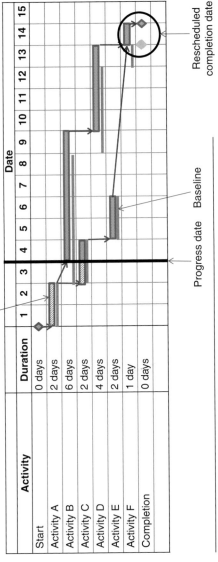

Completed activities are shaded

Activity	Duration
Start	0 days
Activity A	2 days
Activity B	6 days
Activity C	2 days
Activity D	4 days
Activity E	2 days
Activity F	1 day
Completion	0 days

Progress date   Baseline

Rescheduled completion date

Figure 3.3.16 A progressed programme with delay

completed) by the progress date of day 3. Rescheduling of the project has pushed the activities beyond the baseline plan, and the completion date is now one day later than originally planned.

The construction manager will now have to review the programme and available resources to try to catch up the lost time and bring the work back on programme.

In some cases, changes to the work from that planned can relocate the critical path and available float within a programme, meaning that the construction manager needs to focus on a new set of critical activities.

But time management is not simply about managing delays; in some cases works may be found to be ahead of programme, and it may be necessary to re-plan the forthcoming works based on the availability of subcontractors and resources, or different work areas becoming available.

3.3.10 Time management: summary

Time management involves both the planning and control of activities within specified time-frames, and with allocated resources to meet those time-frames.

Construction management production planning should focus not just on the activities individually, but more on how the activities link together to form the project as a whole. Planning must be carried out carefully, and close attention paid to the sometimes complex relationships between construction activities. This should be reflected in the communication of the plan through a clear and easy-to-understand programme.

While durations and resources form necessary parts of construction programmes, it is the relationships and logic among the activities that make programmes useful management tools. By using logic, the sequence, relationships and timing of the relationships among all of the activities can be clearly shown, which in turn identifies the critical path and any float.

This provides useful information for the construction manager, and helps inform the daily decisions they have to make around time and resource management on their site. Knowing which activities need to be kept closely in check and which have some spare time can make the construction manager much more efficient and effective.

The programme is a dynamic construction management tool, and its monitoring enables the construction manager to closely control the site works. If programmes are not reviewed and updated, the construction manager can never be certain whether the project will complete on time or how they should plan future resources. A fully logic-linked programme enables the construction manager to update the plan on a regular basis, and to take corrective action and re-allocate resources as necessary to keep the work on programme and ensure effective time management on the project as a whole.

Discussion point comments

Discussion Point

What other key dates or time periods might need to be included? Think about key dates for specific types of projects; for example, schools.

Other key dates that require inclusion in the programme are fixed holidays; for example, the Christmas break in the UK and other Bank Holidays when the site will be closed. Training periods for new maintenance staff may be required before the building is handed over, and some projects include a time period where the contractor helps the client to move in. In the case of a school this may be during the half-term or summer holiday, and so this date cannot be moved without causing significant problems to children, teachers and parents.

3.4 Cost management

3.4.1 Introduction

This section will focus on cost management, particularly **cost planning and cost control** and **cash flow**.

Cost planning seeks to predict the costs for a project, and both clients and contractors will carry out this process. For the client, this forms their budget for the project. For the contractor, their predicted cost, or **cost forecast**, forms the tender figure they submitted when they won the project.

The function of cost control systems is to monitor the actual project costs and compare them against the predicted costs or budgets. As comparisons are made, gains or losses against budgets will be identified.

Cost management operates at all tiers in the supply chain: clients/employers, consultants, contractors, subcontractors, sub-subcontractors, suppliers and builders' merchants, manufacturers, and raw material producers.

This section focuses on cost management operations in which construction managers are likely to be involved. They are viewed from a main contractor's perspective and therefore also include cashflow as a key consideration. However, some appreciation of the client's methods of cost planning and control are also needed to provide the background against which contractor cost management is carried out.

The remainder of this section will focus on an example project, Project ABC, to illustrate how cost planning and control systems operate in practice. A contractor bids £2,380,820 for the project, with a contract duration of 50 weeks.

3.4.2 Client cost planning

Construction clients and their advisers need to control costs during the design and construction phases of a project, and to try to ensure that projects are completed within the allocated budget.

Clients often employ their own professional quantity surveyors to advise on cost at pre- and post-contract stages; that is, before and after contracts are agreed with contractors. At the first identification of client need, alongside some description of the work required, there will be an early attempt to estimate cost. Indeed, budgets may be the first piece of information created around the project, and the design is developed around the amount of money available.

As designs develop, with better information the predictions of cost can be updated.

Forecasting the client budget

In the early stages, forecasts may simply be in the form of a range. For example, an office development may be in the range of £1,800 to £2,000 per square metre of floor area. These types of rates are available in many commercially available price books, and quantity surveyors will also be able to formulate their own forecasts based on previous experience. As projects move from scheme design to 1:200 scale plans and elevations, the cost of individual elements of the work may be predicted.

The **Building Cost Information Service** (BCIS), operated by the RICS, provides a database of cost information based on real life completed projects. Client quantity surveyors feed cost data into the BCIS database, and in return they gain access to cost data on other projects submitted to the BCIS by their professional colleagues. Cost per square metre of floor area is provided for six elements:

- Substructure
- Superstructure
- Finishes
- Fittings and furnishings
- Services
- External works.

For example, the database may contain details of a large four-bed detached house, with in-situ strip foundations and slab that cost £119.00/m² for a floor area of 200m².

Quantity surveyors looking to estimate the cost of a future project with a floor area of 175m² may estimate the cost at:

$$£119.00/m^2 \times 175m^2 = £20,825$$

BCIS also provides information on **tender price indices** (TPIs) and location factors.

TPIs are records of historic price movements and predictions of future tender price rises or decreases. Predictions of future indices are made by bringing together general market intelligence and also data from government, the Treasury, Bank of England and other eminent forecasting organisations.

The TPI was set at 100 in 1985. When the index reached the level of 110 (say, in 1987), that indicated a 10 per cent tender price increase between 1985 and 1987.

If there were a 10 per cent increase in tender prices between 1987 and 1989, that would be reflected in a TPI of 121, not 120. That is $(121 - 110/110) \times 100 = 10\%$.

The BCIS TPI in the last quarter of 2007 was 251 and the fourth quarter of 2013 it was down to 237f. That represents a 5.5 per cent tender price reduction thus: $(251 - 237/251) \times 100 = 5.5\%$. This reflects the difficulties in markets during this time period.

However, the BCIS forecast TPI for the last quarter of 2018 is 312f, representing a forecast increase between 2013 and 2018 of 24.3 per cent thus: $(312 - 251/251) \times 100 = 24.3\%$.

BCIS also has a cost index, and this rises and falls at a different level to TPIs.

It is important to distinguish between cost and tender indices. **Cost indices** are based on costs being incurred by contractors as the first tier in the supply chain: labour, plant, materials and preliminary costs. Cost indices will rise as costs rise, and fall as costs fall. Ideally, contractors would like to pass on cost increases to clients through increased tender prices; or indeed to give the benefits of any costs that fall.

However, if work is scarce, contractors may decide to absorb increases themselves by bidding with lower profit margins. Therefore cost indices rise, but TPI indices rise more slowly, stay level or even fall. Alternatively, when work is abundant, contractors may only be willing to bid for work with larger profit margins; they are motivated to pull back losses perhaps, from years when work was scarce. Therefore, contractors' costs may be rising, and they are able to pass these rising costs on to clients in tender bids, and also increase their bids further with enhanced profit. Contractors may argue that they are not increasing profit; it is just that when work is scarce they do not price risk on projects for fear of not winning work. When work is more abundant, they are able to price risk and still win work.

Figure 3.4.1 illustrates the TPI and GBCI (General Building Cost Index) for the period 2007 to 2018.

Location factors may be used within indices to take account of the area of the UK in which buildings are constructed. Table 3.4.1 illustrates some BCIS 2014 location factors.

An example of an arithmetic calculation to adjust for location and TPI is as follows:

There is a proposed project in the northwest of England with a location factor of 93 and a project mid-point of quarter 4, 2015, BCIS forecast TPI 256f. The foundation construction comprises in-situ concrete strips, trench block and precast T beams with block infill and screed. A project with a similar specification was completed in the West Midlands, with a location factor of 96 and a project mid-point of quarter 4 2012, BCIS TPI = 225. The substructure elemental rate is £119.00/m².

The rate for the proposed project is as follows:

Location adjustment: $(119/96) \times 93 = 115.28$.

TPI adjustment: $(256 - 225/225) \times 100 = 13.7\%$ *increase*

$115.28 \times 1.137 = £131.07/m^2$

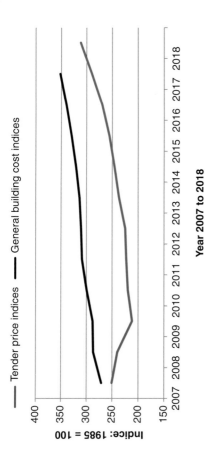

Figure 3.4.1 Line diagram to illustrate tender price and general building cost indices. Note that in the period 2007 to 2010, costs are rising while tender prices are falling. In the period 2015 to 2018, tender prices are forecast to rise more rapidly than costs.

Developing the client budget

Professional quantity surveyors provide clients with predictions of costs up until the point when projects go out to tender, or up until the start of negotiations with the supply chain in non-traditional forms of procurement.

If at any point there is the feeling that likely tender bids will exceed budgets available, design teams will undertake **value engineering** or **value management** exercises; that is to examine individual elements of building designs, and to re-evaluate whether the same

Table 3.4.1 BCIS 2014 location factors

Area	Location factor
Yorkshire and Humberside	96
East Midlands Region	94
East Anglia Region	100
Southeast region	107
Greater London	117
Southwest Region	105
West Midlands	96
Northwest	93
Northern Region	89
Wales	99
Scotland	97
Northern Ireland	60
Islands	136

functionality can be achieved by using different materials, methods or technologies.

At some point prices or budgets are agreed; these become headline figures reported to boards of directors in the private sector, or perhaps to council officials in local authorities.

Client cost control systems then move to ensuring that final build costs do not exceed these 'headline' approved tender figures. The tender figure becomes the budget against which actual costs are monitored.

3.4.3 Contractor cost planning

Companies are in business to make a profit and so their cost planning must be efficient and effective to ensure that actual construction costs are fully forecasted, to allow for the realisation of profit over and above these costs.

When contractors bid for projects, estimators will make forecasts of how much the final cost of projects will be if they are successful in winning the work. That cost will include the following:

- Labour
- Plant
- Materials
- Subcontractors
- Preliminary items.

These elements are considered within the context of the bid for Project ABC. More attention is paid to the cost planning of labour, subcontractors and preliminary items, as these are often most relevant to the construction manager.

Preparing the bid

A typical contractor's bid for Project ABC is shown in Table 3.4.2. The estimate of cost for the main contractor's own work is £190,523, plus £198,300 for preliminaries.

At the final stages of the bid process, the contractors' team will meet in 'tender adjudication meetings'. On larger projects, meetings should involve directors, estimators, planning surveyors and those managers who will be involved in projects if bids are successful. Meetings will appraise estimates of costs and consider whether elements of the estimate are too high and if any competitive advantage has been lost. Alternatively, some parts of bids may be too low, which could leave potentially winning bids with insufficient funds.

There are many high-profile projects in the public domain where major contractors declare losses of millions of pounds on projects, because they underestimated risks at tender stage and did not include sufficient monies in their tenders. In the case of Project ABC, the contractors' team decide to deduct £8,636 from the bid to give it greater competitive advantage. Perhaps they have received some

Table 3.4.2 Contractor's bid; tender summary sheet: Project ABC

		Discount
Main contractor's own work:		
Materials	105,765	5,999
Plant	7,456	
Labour	77,302	
Subtotal	190,523	
Preliminaries	198,300	
Subtotal	388,823	
Late adjustments (deduct)	8,636 → 380,187	
Domestic subcontractors	780,207	20,968
Prime cost and provisional sums		
Nom subcontractors 2.5%	914,500	22,862
Nom suppliers 5%	5,890	295
Prov sums: net	97,309 → 1,017,699	
$1/_{1,115} \times 108,500$		
Subtotal	2,178,093	50,124
Cash discounts: deduct		50,124
Subtotal	2,127,969	
Bonds and project insurances	–	
Firm price addition	7,295	
Subtotal: estimated net cost	2,135,264	
Overheads: 7.5%	160,145	
Profit: 2.5%	53,382	
Risk: 1.5%	32,029	
Total: tender sum to form of tender	**£2,380,820**	

Other tenderers – six contractors: DEF, GHI, JKL, MNO, PQR, STU

late prices from suppliers or subcontractors that are lower than those included in its initial figures.

Main contractors must also consider costs of subcontractors needed to complete the project, and subcontractors must consider the costs of their sub-subcontractors, and so on. Subcontractors often make up a major portion of main contractor bids; in Project ABC £780,207 has been allocated for its own subcontractors and £914,500 for **nominated subcontractors.**

Nominated subcontractors are those selected by the client and their design team, rather than selected by the main contractor. It sometimes occurs in specialist trades like mechanical, electrical or lift installations, where clients have long-standing relationships with companies on other projects and wish to keep those in place. Due to some legal complexities, nomination is not so popular today as it has been in the past.

Nominated suppliers are similarly selected by clients and their design teams. A figure of £108,500 is included as a **provisional sum** within the Project ABC bid. Provisional sums are sums of money included by clients for work that they know they need but are not able to detail at tender stage. The design details and prices are agreed when projects are on site. 11.5 per cent is taken off £108,500, but then added back on at the end for overheads, profit and risk; when prices are agreed for the work to be completed, clients' quantity surveyors will allow contractors to add 11.5 per cent to any net figures in the valuations.

The contractor has been able to secure discounts from its supply chain of £50,124 for Project ABC in much the same way as members of the public are able to gain discounts in the high street. Subcontractors may be commonly asked to provide a 2.5 per cent discount to main contractors.

That discount is supposed to be an incentive to main contractors to pay on time, but as Section 3.4.7 explains, that does not always work well. Often contractors pay their supply chains late, but still keep the discount.

An important point to consider is: who gets the benefit of the discount? In the Project ABC bid, the contractor may wish to submit a competitive price, and therefore deducts the discounts from the tender bid, effectively giving the benefit of discounts to the client. The contractor makes a firm price addition of £7,295. In bid documents, clients often ask the main contractor to take the risk of inflation. The price the main contractor has included for its own work is £388,823, current at the date of tender. The sum of £7,295 is an estimate of likely future cost increases. If actual cost increases are higher, the contractor will gain; if actual increases are lower than this, the contractor will lose.

It is up to tender adjudication meetings to make judgements about lump sums to be added for risk and a percentage for head office overheads; in this case 7.5 per cent has been added which is a common amount. Main contractors whose head office costs are significantly more than this may be at a competitive disadvantage, while those who have lower costs may be at an advantage.

Judgements also need to be made about **profit**. For main contractors a benchmark figure is 2.5 per cent. If work is scarce and contractors need to secure turnover, they may bid at less than 2.5 per cent or even zero profit. In really difficult markets, contractors may 'buy' work just to maintain turnover. They may hope (dangerously) that, 'fingers crossed', losses will not accrue, since sites may be able to build at costs lower than estimated. Buying work is sometimes known as 'suicide' bidding; it has the potential to lead companies into **liquidation**, with serious consequences for all.

If work is abundant, they may judge that they only want projects with good profit margins, and they may submit tenders with higher profit; if that causes them not to win bids, so be it. A sum of 1.5 per cent is added for risk; the contractor may be making a generic judgement about risks on the project, or there may be one or several specific items identified. The contractor has included £32,029 for Project ABC. In some cases that sum of money could be enough to move contractors from a position of lowest bidder to second lowest, which may cost them the project.

Labour costs

Labour is one of the resources over which construction managers have greatest control.

This control stems from time management, where construction managers allocate resources to the programme to ensure that the right labour is equipped with the right plant and right materials to work as effectively and efficiently as possible on site.

There are two elements to consider when weighing up labour costs: the estimators' selection of productivity constants, and rates per hour.

Table 3.4.3 illustrates productivity constants for a variety of trades. Estimators work on the basis of decimal hours per unit of production, since figures in this way are most easily manipulated in spreadsheets. The sum 0.50 hours per m² is therefore used in preference to 2m² per hour.

Table 3.4.4 illustrates a typical 'all-in rate' buildup for a contractor using direct labour. If an estimator judges that a task to be executed by a tradesman will take one hour, a net sum of £14.22 will be included in the tender bid. If the judgement is that a task will take 0.50 hours/m², and there is 1000m², the net sum to be included in the tender will be:

$$0.50 \text{hrs/m}^2 \times 1000\text{m}^2 \times £14.22 \text{ per hour} = £7,110.00.$$

Directly employed labourers have contracts of employment, and employers usually follow the terms and conditions of pay set out in National Working Rule Agreements.

A significant cost for employers when using directly employed labour is the Employers' National Insurance Contribution. The current percentage payment that employers must make to the government is 13.80 per cent of wages. In addition to the basic hourly rate that is paid, contractors must pay overtime rates (perhaps 1.5 hours' pay for every one hour worked on Saturdays), and holiday pay for public and personal holidays. If operatives are on sick leave, contractors will need to include monies to cover for payments made.

Contractors are also required to pay the Construction Industry Training Board a levy on their wage bill, which the Board then uses to fund training, and for grants to contractors to employ apprentices.

Table 3.4.3 Productivity constants

Item	Unit of measure	Plant production constant; hours	Labour production constant; hours
Excavate to reduced level maximum depth not exceeding 250mm	m³	Cat 0.04 hrs/m³	—
Excavate foundation trenches exceeding 300mm wide, not exceeding 1000mm deep	m³	JCB 0.25 hrs/m³	0.25 hrs/m³
Plain concrete beds/slabs 100–150mm thick	m³	—	1.50 hrs/m³
A142 mesh reinforcement in ground slabs	m²	—	0.03 hrs/m²
102.5mm common bricks in 1:2:9 gauged mortar	m²	—	1.25 hrs/m²
100mm thermalite block size 440 × 215, in 1:2:9 gauged mortar	m²	—	0.65 hrs/m²
Proprietary roof trusses, 22.5 degree pitch 5000mm span	Nr	—	1.20 hrs/each
9mm plasterboard to ceilings, over 300mm wide	m²	—	0.24 hrs/m²
75mm lightweight floor screed	m²	—	0.37 hrs/m²
One coat emulsion paint to walls internally, over 300mm wide	m²	—	0.13 hrs/m²
Excavate trenches to receive service pipe, 500mm wide, 750mm deep, including grading bottoms and backfilling	m	JCB 0.09 hrs/m	0.22 hrs/m
900 × 600 × 50mm pre-cast concrete paving flags	m²	—	0.25 hrs/m²

Other overhead-type charges that contractors include in their all-in rate are severance pay and insurances. The total cost of employing a tradesman is shown in Table 3.4.4 to be £26,396.53 per annum, and to calculate the hourly rate this is divided by the total number of hours worked per year: 1856 = £14.42.

Therefore, rather than use directly employed operatives, there are significant incentives for contractors to use self-employed labour. There are, however, strict rules put in place by government; for example, self-employed people must not be employed regularly by the same employer, and may be required to provide their own materials or plant. Self-employed people who develop a regular and continuous work pattern with the same contractor may need to be re-classified as directly employed.

Key incentives for contractors to use self-employed labour are as follows:

- Contractors do not have to pay the 13.80 per cent Employers' National Insurance Contributions.

Table 3.4.4 Labour all-in rate

	Hours /days	£ rate	£ total
Flat time	1893.8 hours	10.46	19,809.15
Non-productive overtime	65.5 hours	10.46	685.13
Public holidays	63.0 hours	10.46	658.98
Holiday pay allowance	176 hours	10.46	1,840.96
Sick pay	5.0 days	21.43	107.15
Subtotal			23,101.37
Employer's national insurance contribution			2,154.64
Training allowance 0.5% of PAYE			115.51
Subtotal			25,371.52
Severance pay and other statutory costs 2%			507.43
Employer's liability insurance 2%			517.58
Total cost of 1856 productive hours			26,396.53
Total labour cost per hour			14.22

Source: BCIS 2013.

- Wages are not usually paid during inclement weather.
- On occasions when contractors need to reduce labour on sites, they can release the self-employed without periods of notice and redundancy pay.
- Responsibility to pay for holidays, travel, tools and income tax lies with the self-employed.

Since employers have fewer overhead costs to bear, and the self-employed have to carry more of their own costs and risks, rates of pay are usually higher than those for directly employed labour.

Self-employed labour is often paid on **price** or **piecework**; operatives are paid for the amount of work they produce, for example, the number of bricks they lay or m² of ceiling they fit. Alternatively, an agreed fixed hourly or '**daywork**' rate can be established; however, this is often thought to reduce productivity as the operatives no longer need to work as fast to receive their daily pay.

Subcontractor costs

The total price included in the contractors' tender is taken as £1,642,241. This is calculated as the figures for domestic and nominated subcontractors of £780,207 plus £914,500 respectively = a total of £1,694,707; discounts of £20,968 and £22,862 are deducted together with the late adjustment sum of £8,636 which is assumed to arise from a more competitive subcontractor quote. Total deductions of £52,466 taken from £1,694,707 give a total of £1,642,241. Having secured these prices for subcontract work at tender stage, the ideal situation is that when the contractor is told its bid is the winning bid, it should simply place orders with subcontractors at the values they have bid. However, many contractors invite a second round of bidding from their subcontractors. The first round is a 'bid for a bid'; the contractor is bidding for a project and is inviting subcontractors to form part of that bid team. Statistically, if six main contractors are bidding to clients, there is only a one in six chance of succeeding. The second round of bidding, assuming the one in six chance has 'come off' for the main contractor, is a 'bid for an order'.

Contractors now ask for subcontractors' 'best price' and secure this by a process sometimes known as a **dutch auction**: contractors barter with subcontractors in an attempt to force prices down. Second-round bidding and dutch auctions are often frowned upon by industry leaders. It is perceived that the bidding and re-bidding add unnecessary costs to our processes, and that subcontractors inflate their first-round bids, since they know that the main contractors will come back for lower prices in round two. Ultimately, construction clients end up paying for this waste.

Dutch auctions can work both ways. In buoyant markets, main contractors may lose out if they delay in placing subcontract orders. For example, a main contractor may wait until it is many months into a project to place an order for a floor finishes package. They may approach a specialist and ask for a lower price than that submitted in round one; the reply may be, 'I am sorry I am having to increase my prices since I am very busy, and my new price to you is now x per cent higher than my round one bid'.

Preliminary items

Also known as preliminaries, or just 'prelims', these are costs incurred by a main contractor on site that are not directly associated with specific elements of the measured works. Main contractor staff salaries are often the largest part of preliminary costs, but they are also needed for site management and may include temporary accommodation, temporary roads, scaffolding and tower cranes.

When pricing preliminaries, contractors allow for 'fixed' and 'time-related' costs. Typically, fixed costs are one-off charges that may arise at any time on projects (e.g. temporary service connections at the beginning of projects and disconnections at the end, or delivery and collection of site cabins). Time-related charges may be staff salaries, the hire of site cabins or the hire of tower cranes.

Construction managers need to look very closely at preliminary budgets to ensure that they are suitable for the project they are being asked to manage.

Preliminaries may account for between around 6 per cent and 15 per cent of the total cost of a project. On a simple project, such as a steel-framed factory development at the lower end of that range, and for more complex projects, perhaps hospitals that will need more staffing, at the higher end.

Ideally, construction managers would like to have generous preliminary budgets, but in the real competitive world, if contractors' bids to clients contain generous sums of money, then contractors may not win projects. Preliminary budgets must, however, contain enough money to maintain standards for welfare facilities and to keep people safe. The quality of temporary site establishments on some projects is excellent or outstanding: showers, restaurants, carpets in meeting rooms, car park spaces, storage sheds, computing facilities. On some occasions, clients specify minimum requirements in their bid documents.

A really important element for any construction manager is the amount of staff support they are able to call upon. The budgets may allow construction managers to have teams of professionals working on their projects, including setting out engineers, planning surveyors, quantity surveyors, supervisors and foremen. Alternatively, if there is little money in the budget, construction managers may need to take a more active part in some of the work normally completed by these people. Construction managers may also wish to employ general labourers to keep sites clean and safe, but may be under pressure not to do so if provision has not been made in the budget.

On some projects the quality of site accommodation may be poor and too little support is available from other staff for construction managers. Some contractors may argue that if they allowed room in their tenders for better accommodation, or for better support to construction managers, they would not win work. Construction managers may welcome the opportunity to work on projects where the preliminary budget is appropriate, but decline to work on projects where budgets are inadequate.

Construction managers 'can only do what they can do'; but if they are forced to undertake many tasks themselves because they do not have budgets for support staff to help, many things on sites may suffer, including the production parameters of time, cost, quality, health and safety, and environmental management.

3.4.4 Client cost control

It can be inevitable that things on site will not progress exactly as planned. While every attempt should be made to minimise change, often it does occur, and it does involve cost.

Which party to the contract should pay extra costs depends on the nature of the change and the nature of contracts that have been agreed. For example, if ground conditions are worse than envisaged, in some cases contracts are set and agreed so that the client takes this risk, and will pay the contractor for extra excavation, filling or concrete. In other contracts, the contractors are asked to take this risk when bidding; they have the opportunity to add a sum of money to their tenders for the risk of bad ground. If the ground is not worse than envisaged the successful contractor can keep the money; but there is also the possibility that extra costs in the ground may exceed the risk sum allocated, and the contractor may lose money.

One area where the client may have to pay extra costs is if they change their mind about the design. That may involve taking down and replacing something that has already been built, or changing a

3

specification or adding some extra work. As things that change affect cost, they are called 'variations', and they are 'valued' or priced fairly in accordance with the contracts.

The data in Table 3.4.5 illustrate how the client may control costs on Project ABC as work proceeds. The project has been running for four months and the contractor has been paid £712,656, including £25,000 for variations. The forecast final account is the same as the tender figure: £2,380,820.

3.4.5 Contractor cost control

Cost control systems should draw immediate attention to operations that are being carried out at uneconomic rates. The primary reason for

Table 3.4.5a Clients' cost control report: Project ABC Valuation No. 4

	Contractor's B of Q	Paid at this valuation	Variations paid at this valuation	Total paid at this valuation	Estimated final account
Preliminaries	301,785	102,600	15,000		
Groundwork	90,000	68,000	2,000		
In-situ concrete	72,000	50,000	2,000		
Masonry	96,645	60,000			
Structural	80,000	50,000			
Linings	75,000	15,000			
Windows/doors/stairs	90,000	16,000			
Surface finishes	110,000	–			
Furniture/equipment	65,000	–			
Building fabric sundries	100,000	–			
Paving/planting/fencing	80,000	52,000	6,000		
Disposal systems	74,500	44,656			
Piped supply systems	5,000	–			
Mechanical heating	2,000	–			
Electrical supply	10,000	–			
Provisional sums	108,500	–			
Nominated subcontractors	914,500	200,000			
Nominated suppliers	5,890	–			
Dayworks/variations/ext of time	–	29,400	–	–	–
TOTAL	2,380,820	687,656	25,000	712,656	2,380,820

Table 3.4.5b Clients' cost control report: Project ABC Valuation No. 4

Month number	Project start forecast net valuations after retention	Actual payments	Re-forecast payments
1	97,000	116,400	-
2	213,400	256,080	-
3	460,750	551,782	-
4	577,344	691,277	
5	853,600		
6	1,154,697		
7	1,455,000		
8	1,732,032		
9	1,920,600		
10	2,056,400		
11	2,182,500		
12	2,321,300		2,321,300
18	2,380,820		2,380,820

Contract period in weeks	50	Percentage period expired: 34%
Extension of time	–	Gross valuation as % of tender sum: 29.9%
Number of weeks in progress	17	Retention held: £ 21,379
Number of weeks to run	33	
Number of weeks ahead/behind	2	

undertaking cost control is the early identification of practical problems that need attention and corrective action; inefficiencies in production should be quickly corrected. The longer operations are allowed to

Discussion Point

The client is concerned that £25,000 has been paid to the contractor for variations. The client has not changed the 'brief', which was 'signed-off'. How would you advise the client?

Discussion Point

The client is concerned that the contractor is preparing an extension of time claim, to include loss and expense, because of ground conditions being worse than anticipated and because of variations. How would you advise the client?

3.19

Discussion Point

The client is concerned that the actual payments made as a result of the monthly valuations are higher than forecast. How would you advise the client?

3.20

Discussion Point

The client is concerned that the forecast final account figure is the same as the tender sum. In the light of the problems on the project, how would you advise the client?

continue at uneconomic rates, the greater the risk of financial damage. It is therefore of the utmost importance that cost control systems are quick, reliable and accurate, and that the construction manager is able to use them to their full extent.

Cost control systems should also be able to provide feedback to the estimators who are responsible for pricing tenders. If rates are too 'tight' then even effective and efficient work can lose money, and so they must be adjusted on the next project to avoid it happening again. If rates are too 'generous' and sites are making more money than anticipated, this could mean uncompetitive tenders. Construction managers should provide feedback to their estimating teams at head office to ensure that the company information is accurate and up to date.

Cost control systems can generate reports on the following time basis:

- Weekly – completed by the end of the week, following the week of production.
- Monthly – completed by the end of the month, following the month of production.
- Completion of projects.

Where possible, cost reports should be produced on a weekly basis, and compiled the week after the actual performance. Many activities in construction may last for a relatively short period. Consider brickwork lasting eight/nine weeks or two months. A weekly report available at the end of week 2, showing losses, gives the opportunity for corrective action. A monthly report available at the end of month 2, showing losses, does not give the opportunity for corrective action; it's too late.

Cost control systems are often founded on coding systems.

Invoices are given numerical codes to enable all similar coded costs to be collated. For example, a material invoice may be for concrete used in a temporary tower crane base, in which case it will be allocated a preliminary code. Alternatively, an invoice for concrete poured into foundations to the permanent building will be allocated a code related to production.

Net gains and net losses

Contracting systems are often set to measure **net gains**.
Consider the following question:

Assuming a profit has been included in a bid, who does the profit belong to? To the site, or to head office? To illustrate this further, consider the position of a project that is estimated to cost £90k. Ignore for this exercise head office overheads. It is a buoyant market; the contractor adds £10k for profit; that is 11.11 per cent on top of the estimated cost of £90k to give a bid price of £100k.

The contractor wins the project; the final account payment is the same as the tender bid, at £100k. Scenario 1: the actual cost of the project when completed is £85k. Question: does this constitute a £5k or £15k profit? Scenario 2: the actual cost of the project when completed is £95k. Question: does this constitute a £5k profit or a £5k loss?

The net gain is the sum achieved below the estimated net cost; thus in scenario 1 an actual cost on completion of £85k would represent a £5k net gain. In scenario 2, an actual cost on completion of £95k, though bringing £5k profit into the company, actually represents a £5k net loss. Therefore it may be argued that the £10k added to the bid for profit belongs to head office. Construction managers are charged with making net gains; cost reports are often based on net figures (the £90k), not the gross figure of £100k.

Contractors are aspiring continually to net gains: net gains, net gains and net gains. Large UK main contractors announce proudly in the press if they make 4 per cent profit on turnover; they may have bid projects generally at 2.5 per cent, but through efficient working they have squeezed out net gains to secure a further 1.5 per cent and an overall 4 per cent. Working from the net, a project that achieves £15k profit on £85k costs is an excellent percentage profit margin of $15/85 \times 100 = 17.64\%$.

When things are going well on projects, it is important that contractors secure net gains; they should be building at lower than estimated cost. The nature of construction work is that while things may be going well now, around the corner on all projects there are potential net losses; for example, bad weather, an element of work where risk had been underestimated or a delay caused by labour shortages.

Hopefully over the life of one project, net gains on some trades may outweigh net losses on other trades. Some projects will lose money: not just a net loss where income is £100k and cost is £95k; but income of £100k and cost of £105k or more. Hopefully over the life of one contractor's financial year, gains on most projects will outweigh losses on others; and thus contractors will report an overall end-of-year profit. However, taking into account trading on all projects, there are many examples in the UK where main contractors report end-of-year losses.

In some cases these losses are so high that they cannot be counterbalanced by profits in previous years, or projections of profits in the future. As a consequence, on some occasions contractors may be sold, or go into liquidation with potentially serious consequences for jobs in those organisations and in their supply chains.

In the case of Project ABC, the construction manager is responsible for a budget of £2,135,264 plus the risk allocation of £32,028: a total of £2,167,292. The balance of the money (£2,380,820 – £2,167,292 = £213,528) belongs to head office. Construction managers and their teams are 'charged' with making net gains on their projects.

Weekly cost control for labour and plant

For contractors, weekly cost reports offer an effective method of cost control.

Trends will be highlighted within a matter of days following the completion of work. Weekly reports are particularly useful in highlighting extra work carried out for clients for which extra payment should be sought, or work executed which may be chargeable to, say, subcontractors or suppliers.

The basis of weekly labour and plant cost reports are accurate records of work completed, usually in the form of allocation sheets. These should identify the persons involved, the work done and the time spent on each operation. An allocation sheet for a gang of groundworkers is illustrated in Table 3.4.6.

Similar to self-employed people, often payments made to directly employed operatives are on 'price' or 'piecework'. If there is a good, productive week, perhaps when the weather is fair, operatives may have the opportunity to earn good money. If the weather is not so good, earnings may be lower.

Table 3.4.7 illustrates a cost report for a directly employed gang of bricklayers paid on 'price' or 'piecework'. The cost side is the wages paid to the bricklayers. A notional 25 per cent is added to the wages paid to cover contractors' overheads as described above. The budget side is the amount the contractor will be paid, based on rates it has included in Bills of Quantities. The cost report shows a net gain of £1,203.65.

Alternatively, the basis of payment may be fixed hourly rates, commonly called 'daywork', and thus tradespeople receive the same weekly wage irrespective of higher or lower productivity.

If delays result because of weather or perhaps other factors such as material shortages or design information not being available, operatives will be paid the same amount of money. Table 3.4.8 illustrates the contractor's costs when work is completed on the basis of a fixed hourly rate and is again compared to the contractor's budget. This report shows a net gain of £953.49. Payment to tradespeople on a fixed hourly rate may be on projects where work is difficult to measure or predict times accurately, such as refurbishment, alterations or maintenance contracts.

Table 3.4.6 Labour and plant allocation sheet

	Mon	Tues	Wed	Thurs	Fri	Sat	Sun	Total
Labour								
Miles	8	8	8	8	7			39
Mason	8	8	8	8	7			39
Miller	8	8	8	–	7			31
Total	**24**	**24**	**24**	**16**	**21**			**109**
Plant								
JCB3 inc. driver	8	8	8	8	7			39
Labour tasks								
Excavate foundation trenches	10	10	9		8			37
Concrete	6	6	6		3			21
Backfilling, spread and level	1	1	1	1				4
Transporting materials; double handle	4							4
Assisting engineer with levels and setting out	1	2	2	1	3			9
Excavate and backfill service trench				4				4
Cleaning site compound		2						2
Unloading materials	2				4			6
Inclement weather		3	6	10	3			22
Total	**24**	**24**	**24**	**16**	**21**			**109**
Plant tasks								
Excavate foundation trenches	3	3	3		2			11
Concrete	1	1	1		1			4
Backfilling, spread and level		2	4	2	2			10
Transporting materials; double handle	4			2				6
Excavate and backfill service trench				4				4
Cleaning site compound		2						2
Inclement weather					2			2
Total	**8**	**8**	**8**	**8**	**7**			**39**

Table 3.4.7 Weekly cost reconciliation: directly employed 2 +1 bricklayers paid on rates or piecework

Names: Smith, Jones and Brown
Date: Week commencing 01.04.201x

| Item | Cost to the contractor | | | | Budget; money received by the contractor | | | | |
	Gang hours; 3 men	Piecework rate	Wages	25% addition to wages for contractor's overheads	Quantity	Unit	Rate	Amount	Net gain or loss £
1/2B facings	5000 No	320.00 /1000	1600.00	2000.00	84.74 (5000 No. /59 bricks per m^2)	m^2	36.60	3,101.48	1,101.48
1/2B facing brick soldiers	12m	6.90	82.80	103.50	12	m	8.80	105.60	2.10
140mm thermalite blockwork	50m^2	9.80	490.00	612.50	50	m^2	16.00	800.00	187.50
1200mm lintels	2 No	2.00	4.00	5.00	2	No	2.06	4.12	−0.88
2400mm lintels	4 No	5.00	20.00	25.00	4	No	3.57	14.28	−10.72
Take down and rebuild brickwork; wrong dimension on drawing	9 man hours	10.46	94.14	117.67	Assume costs will be recovered from the client			117.67	0
Setting up new place of work on west wing	6 man hours	10.46	62.76	78.45				nil	−78.45
Totals			2353.70	2942.12				4148.87	1201.03

Table 3.4.8 Weekly cost reconciliation: directly employed 2 +1 bricklayers paid on a fixed rate of pay

Names: Baker, Butcher and Barnes
Date: Week commencing 01.04.201x

Name	Cost to the contractor; wages				Budget; money received by the contractor					Net gain or loss £
	Hours	Rate	Wage	25% addition to wages for contractor's overheads	Item	Quantity	Unit	Rate	Amount	
Smith	39	10.46 plus 5.43 spot bonus = 15.89	619.71	774.63	1/2B facings	59.32 (3500 No. / 59 bricks per m²)	m²	36.60	2171.47	
Jones	39	15.89	619.71	774.63	1/2B facing brick soldiers	12m	m	8.80	105.60	
Brown	39	7.87 plus 3.94 spot bonus = 11.81	460.59	575.73	140mm thermalite blockwork	40m²	m²	16.00	640.00	
					1200mm lintels	2 No	No	2.06	4.12	
					2400mm lintels	4 No	No	3.57	14.28	
					Take down and rebuild brickwork; wrong dimension on drawing	9 man hours x 15.89	Assume costs will be recovered from the client		143.01	
					Setting up new place of work on west wing	6 man hours			nil	
Totals			1700.01	2124.99					3078.48	953.49

You will note that the productivity levels in Table 3.4.7 are higher than those in Table 3.4.8: 5000 bricks laid versus 3500 and 50m² of blocks laid versus 40m². This may be because the gang represented in Table 3.4.7 are motivated to work more quickly since they are on price or piecework, whereas the gang represented in Table 3.4.8. are on daywork.

Table 3.4.9 illustrates a cost report for a gang of groundworkers with a machine, working on a self-employed daywork basis. There are no overhead charges to add to the amount of wages paid. This report shows a net loss of £42.14.

Weekly cost reports are often time consuming and expensive to prepare. For some construction companies, weekly systems may be considered impractical due to the cost of staff time to produce reports, and management time to consider them. In such cases cost control systems may only be carried out on a monthly or end-of-project basis. Some construction managers are uncomfortable with weekly cost reports, and see them as a method used by senior managers to pry into the detailed running of sites.

With weekly cost reporting systems, construction managers will be aware that they need to be very vigilant, since it is effectively their personal performance that is being monitored. Without weekly cost reporting systems, it may be argued that some construction managers will be less vigilant, and possibly feel they can get away more easily with any mistakes they make.

Weekly control of materials and reconciliations of waste

Since weekly cost reports for materials can be difficult to produce, an alternative method of materials control is weekly **reconciliations** of waste. There are two elements to consider in whether net gains or net losses are incurred on materials:

1 The rate used in the tender versus actual purchase rate (e.g. concrete 28N/mm³, tender rate £92.13/m³, actual purchase rate £89.50/m³).

2 The percentage waste rate used in the tender versus actual waste (e.g. concrete in foundations tender waste 5 per cent, actual waste rate 7 per cent).

Construction managers may have little control over whether there are savings or losses on material money purchase rates; construction managers do, however, have total control over materials wastage rates.

Materials purchased at a lower rate than in the tender may be called a 'buying gain'. A buying gain may offset in money terms any wastage that is higher than the bid waste value. Hopefully, monthly cost reports will show buying gains and material waste as less than the bid waste value.

The percentage waste rate can be calculated through reconciliation. Table 3.4.10 indicates a reconciliation of the use of ready-mixed

Table 3.4.9 Weekly cost reconciliation: self-employed groundworkers paid on daywork

Names: Miles, Mason, and Miller.
Date: Week commencing 01.04.201x

Item	Cost to the contractor; wages			Budget; money received by the contractor				Net gain or loss (£)
	Hours	Rate	Money	Quantity	Unit	Rate	Amount	
Labour								
Excavate foundation trenches	37 hrs	12.00	444.00	98	m³	2.57	251.86	−192.14
Concrete	21 hrs	12.00	252.00	25	m³	13.35	333.75	−81.75
Backfilling, spread and level	4 hrs	12.00	48.00	60	m³	4.42	265.20	217.2
Transporting materials; double handle (take money from preliminaries as budget)	4 hrs	12.00	48.00	4	hrs	12.00	48.00	0
Assisting engineer with levels and setting out (take money from preliminaries as budget)	9 hrs	12.00	108.00	9	hrs	12.00	108.00	0
Excavate and backfill service trench	4 hrs	12.00	48.00	75	m	2.50	187.50	138.50
Cleaning site compound (take money from preliminaries as budget)	2 hrs	12.00	24.00	2	hrs	12.00	24.00	0
Unloading materials (take money from preliminaries as budget)	6 hrs	12.00	72.00	6	hrs	12.00	72.00	0
Inclement weather	22 hrs	12.00	264.00	22	hrs	0	0	−264.00
Total	**109 hrs**		**1308.00**				**1290.31**	**−17.69**

Table 3.4.9 (Continued)

Names: Miles, Mason, and Miller.
Date: Week commencing 01.04.201x

Item	Cost to the contractor; wages			Budget; money received by the contractor				Net gain or loss (£)
	Hours	Rate	Money	Quantity	Unit	Rate	Amount	
Plant JCB3								
Excavate foundation trenches	11 hrs	25.00	275.00	98	m³	7.20	705.60	430.60
Concrete	4 hrs	25.00	100.00	25		0	0	−100.00
Backfilling, spread and level	10 hrs	25.00	250.00	60	m³	0.67	40.20	209.80
Transporting materials; double handle (assume money not available in preliminaries as budget)	6 hrs	25.00	150.00	6	hrs	0	0	−150.00
Excavate and backfill service trench	4 hrs	25.00	100.00	75	m	2.73	204.75	104.75
Cleaning site compound (assume money not available in preliminaries as budget)	2 hrs	25.00	50.00	2	hrs	0	0	−50.00
Inclement weather	2 hrs	25.00	50.00	2	hrs	0	0	−50.00
Total	**39**		**975.00**				**950.55**	**−24.45**
Overall total			**2283.00**				**2240.86**	**−42.14**

concrete. It determines the amount of waste overall and converts that to a financial amount.

In the case of the foundation trenches, the reconciliation shows actual waste far in excess of allowances made in the tender. Controlling waste in foundations can be difficult, since machine buckets cannot excavate with absolute precision, and with unstable ground or bad weather the sides of trenches may collapse as excavation proceeds.

Monthly cost control systems

On larger contracts spread over a period of several months, it is usual to implement cost control systems that run in conjunction with monthly **interim valuations.**

Table 3.4.11 illustrates a typical cost control document for Project ABC. It is at valuation No. 4.

The 'Total construction budget' of £2,380,820 is based on the figures submitted to the client in the tender bid.

The budget is split into the major elements of labour, plant, materials, subcontractors, preliminaries and provisional sums.

The 'Cost to date' figure of £628,176 can be provided by head office. The 'Construction budget to date' determined by the contractor's quantity surveyor of £639,156 shows net losses for all elements, but in overall terms a gain of £10,980.

At this stage, if things had progressed as planned, the project should have been making a profit of £22,930. However, there are losses of £11,116 and £834 on company overheads: a total of £11,950.

Subtracting this figure from £22,980 reconciles a gain of £10,980; that is, £11,950 less than hoped for.

Contractors will always be looking ahead towards the end of projects and making forecasts about what profits will be on completion. In this case, since the contractor is four months into the project, it is able to make more accurate predictions of final cost than were made at tender stage. Here, the contractor's quantity surveyor is predicting a profit of £202,733, compared to the estimated profit at tender stage of £85,416; this represents a net gain of (£202,733 − £85,416) = £117,317.

The report shows that the contractor has 'over-claimed' £44,100.

3.21

Discussion Point

How does the contractor propose to secure this net gain? What is the consequence for the construction process of seeking such gains?

Table 3.4.10 Weekly materials reconciliation for in-situ concrete

Day	Weekly pour number	Location of pour including gridline (GL)	Method of pour	A Amount of concrete delivered: m³	B Amount of concrete that should have been used measured from drawings: m³	C Waste allowed in tender (%)	D = A–B/B Actual waste (%)	E = B x C Actual amount paid for including tender waste allowance (m³)	F = A–E Gain or loss (m³)
Monday	1	Foundation trenches west: GL18-20	Direct discharge	28	24.5	5	14.3	25.7	−2.3
Tuesday	2	Podium slab GL10-11	Crane and skip	16	15.8	2.5	1.2	16.2	+0.2
Wednesday	3	Foundation trenches west: GL20-22	Direct discharge	26	23.1	5	12.6	24.3	−1.7
Wednesday	4	Columns ground to 1st floor: GL 1-4	Crane and skip	6	5.9	2.5	1.7	6.0	0
Thursday	5	Podium slab GL11-12	Crane and skip	22	21.1	2.5	4.2	21.6	−0.4
Thursday	6	Foundation trenches west: GL22-24	Direct discharge	15	13.1	5	14.5	13.8	−1.2
Friday	7	Columns ground to 1st floor: GL 5-6	Crane and skip	5	4.8	2.5	4.2	4.9	−0.1
Totals				118	108.3	-	-	112.5	5.5

Finance: ~~Gain~~/loss 5.5 m³ @ circa £90.00/m³ = £405.00

Table 3.4.11 Main contractor's monthly cost and valuation statement

Contract period in weeks	50	Number of weeks in progress	17
Extension of time	0	Number of weeks to run	33
Tender period	50	Number of weeks ahead/behind	2
		Percentage period expired	**34%**

Comparison of values up to **30.04.2015**
Certificate number ..**4**... with contract costs and construction budget up to this date

Total contract value 'A'	2,380,820	QS gross valuation to date 'D'	712 656
Less: contingencies, etc.	nil	Less: main contractor's materials on site	29 400
Balance 'B'	2,380,820	Over/under claim	44 100
Less: subcontractors as tender	1,642,241	Balance 'E'	639 156
Balance 'C'	738,579	Less: subcont cum net value less discount	394 068
		Balance 'F'	245 088
% value complete E/B x 100	**26.8%**	**% main cont work complete F/C x 100**	**33.1%**

Description	Total construction budget	Cost to date	Construction budget to date	Gain/loss	Final cost forecast
Labour	78,461	38,038	37,999	−39	75,000
Plant	7567	3555	3318	−237	7000
Materials	106,947	61,282	61,102	−180	95,000
Subcontractors	1,642,241	397,593	394,068	−3525	1 560,128
Site on costs	201,274	77,682	68,997	−8685	204,000
Provisional sums	98,768	6200	7750	+1550	85 000
Subtotal	2,135,258	584,350	573,234	−11,116	2,026,128
Company on cost	160,146	43,826	42,992	−834	151 959
Finance charges	nil	nil	nil	nil	nil
Profit	85,416	nil	22,930	+22,930	202,733
TOTAL	2,380,820	628,176	639,156	+10,980	2,380,820

Monthly progress		Cash flow reconciliation	
Previous month cumulative value	485,688	Present gross value	712,656
Value this month	153,468	Less retention	21,379
Present cumulative value	639,156	Present net value	691 277
Present cumulative cost	628,176		
Profit to date	10,980	Cumulative cash received	505 067
Previous month cumulative profit	9,180	Last payment ten days early/late	
Profit this month	1800	Cash outstanding	186,210

Comments

Subcontractor's loss being investigated; claim for extension of time to be submitted; problems being experienced with material thefts.

That is where contractors claim for work they have not done: perhaps 50 per cent of the brickwork, when only 40 per cent has been completed. Contractors often do this to improve their cash flow, as will be discussed in Section 3.4.7, but it is important that costs of work completed are not compared with budget figures for more work than has been completed. Therefore, the over-claim of £44,100 and the value of materials on site of £29,400 are deducted from the valuation amount £712,656 to give a budget figure of £628,176 (£712,656 – (£44,100 + £29,400) = £628 176).

One role of clients' professional quantity surveyors is to ensure that contractors do not excessively over-claim; they may, however, be a little relaxed, since clients do hold **retention**, and by the time contractors receive their payment they will most likely have completed any work they have over-claimed for.

One of the main disadvantages of monthly comparisons is the time lag between execution of the works and reports being available for consideration and action, perhaps as long as one month. However, monthly cost control reports are 'powerful' documents. They may be considered by chief executives and boards of directors. They can give an accurate overview of the well-being of a project. Directors may make important decisions on the basis of monthly cost control documents, and those decisions may be about your future as a construction manager, whether those decisions are positive or negative.

It is therefore important that construction managers at site level and above make sure that they understand the cost control systems of the companies they work for. Systems will vary between companies and, as construction managers move between companies, they will need to learn about new systems. But it is very important that construction managers fully understand the ways cost is managed, planned and controlled on their projects. An attitude of 'I leave that to the QS' could be the downfall of construction managers.

Some companies do not allow construction managers or site quantity surveyors to see monthly cost reports. Site quantity surveyors report budget figures to head office, but head office calculates costs and compares them to budgets. Companies may argue that if construction managers or site quantity surveyors see projects that are making huge net gains, they may relax their efforts; it is just for construction managers to concentrate on efficiency and effectiveness and to let head office 'worry' about cost reconciliations.

Subcontractor cost control

In Project ABC the budget for subcontractors is £1,642,241, as noted in Section 3.4.3. With the project at month 4, it is hoped that the final account for all subcontractors added together will be £1,560,128; that will be a net gain of £82,113.

Making judgements about profits on subcontractors as work proceeds is relatively easy.

Take a groundworks package which is negotiated and sublet following the award of the main contract in the amount of £155,270. The amount the contractor has included in its bid for this work is £163,215 (£163,215/£155,270 = 1.0511). The contractor is making a net gain of 5.11 per cent on each £1 of work executed by the subcontractor. In Project ABC, if the subcontractor has completed £100,000-worth of work at valuation No. 4, the main contractor's budget is £105,110, with a net gain of £5,110 to date, and a forecast net gain on completion of the package of £7,945 (£163,215 – £155,270).

There may be a little complexity in subcontractors' accounts if they execute works outside their initial contracts. For example, a groundworks subcontractor may do some work associated with preliminaries (say, a temporary road) for which an extra payment will be due. Or there may be some design changes or variations, the costs and budgets for which would have to be calculated separately.

Preliminary cost control

A preliminary budget and cost appraisal at month 4 for Project ABC is shown in Table 3.4.12.

Contractors need to make careful judgements about how to release budgets for comparison against costs. At month 4, the preliminary costs are £77,682 and the budget released is £68,997; a loss of £8,685. A large part of this loss arises since the contractor has been on site for four months but is two weeks behind programme.

The concept of 'prudence' must be considered. The cost report suggests that a claim for an **extension of time** will be made to the client. What if this claim is rejected? That may mean that if the contractor completes the project two weeks late, many of its time-related costs, such as staff salaries for those two weeks, will be a loss to the contractor. Prudence suggests that if a loss is a possibility it should be included in cost reports now, and not left as a 'nasty' shock until later. In this case the contractor has released only 3.5 months of its staff budget, and has compared this to four months' costs; obviously a loss.

If the extension of time is granted, the budget may be increased. If the extension of time is not granted, no doubt the contractor will try to re-programme in an attempt to still complete the project on time.

3.4.6 End of contract reconciliations as a form of cost control

Smaller contractors, who undertake projects of a relatively short duration period (i.e. three or four months or less), may have cost control systems that are based solely on end-of-contract appraisals or reconciliation. They may have limited staffing resources to undertake detailed cost analysis.

Table 3.4.12 Contractor's Preliminaries: budget and costs

Cost code	Item		Total budget	Planned budget at four months	Actual budget at four months	Cost	Net gain or loss
	Employer's requirements and accommodation	Site cabins	11,000	3740	3000	3621	−621
		Mobilisation	450	450	450	722	−272
		Demobilisation	850				
	Supervision	Construction manager and foreman	52,550	17,680	15,765	18,623	−2,858
		Site engineer	6000	4800	3800	5321	−1521
		Planner and QS	12,500	4250	3750	3816	−66
	General labour	Attendant labour	12,500	4250	3750	4222	−472
		General attendance	5000	1700	1500	923	577
	Site facilities	Administration	9000	3000	3000	2888	112
		Services	10,860	3000	2800	3211	−411
		Mobilisation	3200	3200	3200	4021	−821
	Temporary works	Maintenance	8400	2800	2500	1565	935
		Mobilisation/ removal	4750	4750	4702	4702	0
		Temporary work	3950	1000	1000	958	42
	Mechanical plant	Lifting	900	300	300	623	−323

	Transporting	8000	4000	1800	4443	−2643
	Concreting	100	100	100	652	−552
Non-mechanical plant	Scaffold	9400	6400	6000	4236	1764
	Instruments	1190	1000	900	823	77
	Miscellaneous	5000	1500	1400	1022	378
	Small tools	1000	500	500	1023	−523
Contract conditions	Insurances	20,000	10,000	1300	1300	0
	Bonds	2000	2000	2000	1880	120
Miscellaneous	Winter working	2000				
	Quality assurance	2000	800	800	522	278
	Safety	5200	4000	3300	6333	−3033
	Setting out consumables	500	400	380	232	148
Subtotal		198,300				
Fixed price allowance	Part of 7295	2974	1000	1000	-	1000
Total		201,274	82,880	68,997	77,682	−8685

This does not mean that such contractors operate regardless of cost; they will be constantly trying to work as efficiently and effectively as possible, and to make cost savings as work proceeds. However, the reconciliation of cost versus budget is not undertaken until project completion.

If losses are noted it is clearly too late for that project, but there may be valuable lessons for projects in the future. Some small builders are sometimes quoted as saying that they do not bother with costs until the end of the whole financial year. They merely report to accountants with bank statements and invoices and say, 'Tell me if I have made money, and how much tax I have to pay.' Underpinning this rather relaxed approach is a feeling that money has been made or lost based only on the cash balances in the builder's bank account.

3.4.7 Cash flow

Cash flow is often called 'the lifeblood of the construction industry'.

Cash flow is a different concept from profit, and arguably just as important. Businesses may have excellent 'paper' profits but poor cash flow if money due is not received on time. Profits and cash flow are linked and profits can influence cash flow: poor profits or losses impacting negatively, and good profits impacting positively.

Construction managers at site level may not see reconciliations of cash balances for individual or multiple projects; often these are matters for head office accountants and financial directors. However, it is important that construction managers have an understanding of company policies, actions and motivations, since the consequences may often permeate down to site level.

Companies need to think about their cash balances at the bank in much the same way that we do individually. If companies have many debts – loans on assets – and overdrafts, they are potentially in a weak position. Companies may have agreements with banks for overdrafts, but if these overdrafts are constantly at their limits it may be the case that banks will withdraw these facilities and leave companies in positions where they cannot pay debtors. The first sign of companies with cash flow difficulties is that they are unable to pay weekly paid operatives, suppliers and subcontractors.

Companies may be forced to take actions they would rather not, at the behest of their banks; it may appear that the banks are controlling companies rather than directors being in control. Alternatively, some companies are in the fortunate position of being cash rich. They may not need overdraft facilities and may have few loans secured against assets. Such companies are in control of their own destiny, using banks only to make and receive payments. They are respected by banks and potential investors, and can feel confident and in a powerful position. If cash-rich companies have one or two problem projects, they are able to survive using their cash balances. If cash-poor companies have one or two problem projects, that may be the catalyst to send them into liquidation.

How the construction industry gets paid

Contractors usually receive **interim payments** for work completed according to monthly **valuations**. Valuations are agreements between clients and contractors' quantity surveyors about the 'value' or the 'worth' of work completed. Contractors are paid for the amount of work they do. If they are behind programme they will receive less money each month on account. If they are ahead of programme they will receive money more quickly; that is until work is completed and final payments are made.

Clients often hold retention from each payment (i.e. a sum of money in case defects arise in completed work). Common percentage levels for retentions are 3 per cent or 5 per cent, although many people argue that retentions are actually not necessary, and advocate zero retention. If retention is held, it is normal to release half of the money held when projects are completed, and the other after the **rectification or defects liability period**, which is usually six or twelve months after project completion. Since clients hold retention from contractors, contractors also hold retention from subcontractors, though not from suppliers.

Contractors and their supply chains need regular payments to keep businesses running. Monthly valuations are largely undertaken at the end of the month; though it may be mid-month or some other arrangement to suit individual clients. Contractors often receive payment around 21 days into the following month (e.g. a valuation completed at the end of January would lead to a payment on or about 21 February).

As contractors receive their payments around the 21st of each month, they are then able to pay their supply chains seven to ten days later, or at the end of the month. An off-quoted payment term in contracts is 28 or 30 days. That is 28 or 30 days after submission of invoices. The valuations that contractors' quantity surveyors agree with clients' quantity surveyors will include subcontractors' invoices submitted at the end of each month. Thus, at the end of January, subcontractor invoices are issued for work completed during that month and payment is made from the main contractor to the subcontractor at the end of February using the monies received on the 21st of that month.

Material suppliers are often paid similarly; all materials supplied to projects in January are invoiced at the end of January, and contractors make payment at the end of February. Plant hire companies that provide mechanical or non-mechanical plant also have the same arrangements.

Contractors have to pay several parties on sites before they get paid; weekly paid operatives will want wages at the end of week 2 on projects, and staff will require their salaries at the end of month 1. Some companies may want payments before they start work; for example, statutory undertakings may require that their invoices are paid before they will make temporary water or electrical connections. Contractors may finance these payments from agreed overdrafts or from surpluses held from other projects.

Potential problems with payments

Clients often pay main contractors on time; there is little attention in the public domain to client late payment. Some contractors pay their supply chains exactly on time, and develop excellent reputations for doing so. However, that is not the case with all contractors, and paying supply chains late is a recurring problem, and the subject of frequent debate. The document 'Industrial Strategy: Government and industry in partnership Construction 2025' gives as one of its visions that 'Construction in 2025 is no longer characterised, as it once was, by [...] late payment'.

Contractors may use two ways to hold on to money they receive from clients for as long as possible.

First, they may legitimately negotiate extended payment terms with their supply chains; perhaps 35 days, 45 days, 60 days, 90 days and even 120 days. It is argued that supply chains are aware of these payment arrangements when they bid, and have the opportunity of including in their prices any finance charges they may incur in waiting for payment.

Second, contractors may agree contracts to make payments within certain periods, perhaps 30 days, but when the due date arrives, payment is not made. Perhaps payment will be made seven or 14 or 30 or 60 days later? While there are methods within contracts to enforce payments, subcontractors and suppliers may be fearful that if they push main contractors too hard, they will not be offered future projects. Some subcontractors get really upset by late payment; others accept it as 'just the way it is'.

From the perspective of construction managers, it may be the case that they request deliveries by telephone for materials such as concrete, bricks or mortar; but delivery is refused because invoices have not been paid. Construction managers need to appreciate that this is a perfectly reasonable position for suppliers to adopt, and make sure they direct this problem back to the person responsible at head office.

Construction managers may also have requests to subcontractors to provide extra labour refused, or note that subcontractors are taking labour away from sites and placing it on projects with main contractors that pay on time; again often a reasonable position from the perspective of subcontractors. While subcontractors have a commitment to meet their programme commitments, they will of course place their labour with main contractors who pay promptly.

Head office accountants will argue that construction managers need to appreciate that late payment is a vehicle that companies use to build up their cash balances and gain respect in the marketplace. Companies need to minimise their reliance on bank overdrafts, since these are expensive and make companies vulnerable to failure.

Construction managers will argue that head office accountants need to appreciate the difficulties that late payments cause on sites. The nature of construction is that mini-urgent crises often develop, and construction managers may need extra help from their supply chain to resolve matters; a last-minute material delivery may be required, or extra subcontract labour. The supply chain may be less inclined to respond if it is not being paid on time. Late payment can lead to delays on site, and construction managers having to reschedule work. Construction managers may argue that these delays cost more than the benefits that accrue from holding on to cash.

3.4.8 Cost management: summary

Cost control systems are important tools to be used by all parties in construction supply chains. It is important that construction managers engage with these systems, since they form the basis for many strategic and operational decisions. Construction managers who do not engage will be disadvantaged at decision-making tables.

The timeliness of cost reports is important; prompt information allows prompt action to be taken before potential losses escalate. Accurate cost data provide the basis for estimating for future projects. Cost control systems can be complex and need extensive time commitment from professional teams; thus they may be expensive to operate. Some companies may commit fewer resources to costing systems than others.

Discussion point comments

3.17

Discussion Point

The client is concerned that £25,000 has been paid to the contractor for variations. The client has not changed the 'brief', which was 'signed-off'. How would you advise the client?

Variations can occur frequently; that is the nature of construction. The design team may find some discrepancies between the specification, Bills of Quantities and drawings. Some of these may result from mistakes made by the design team, or there may be changes in specifications of modern materials or to regulations. Relatively minor variations of this nature are normally absorbed in the contract sum. If the client wants to be particularly vigilant it may ask the design team for detailed substantiation of each part of the £25k.

3.18

Discussion Point

The client is concerned that the contractor is preparing an extension of time claim, to include loss and expense, because of ground conditions being worse than anticipated and because of variations. How would you advise the client?

Contracts often allow claims for extension of time due to ground conditions; clients take the risk. Ground conditions are one part of the construction process that holds the highest risks. The client may ask the design team to refer back to the original ground investigation report. How many boreholes were commissioned as part of the survey? Was it just bad luck that a borehole was not drilled in the vicinity of one patch of bad ground?

3.19

Discussion Point

The client is concerned that the actual payments made as a result of the monthly valuations are higher than forecast. How would you advise the client?

Higher payments at early stages should not cause too much concern, though the client's quantity surveyor does need to be sure that the contractor is not excessively over-claiming. Higher payments should normally indicate good progress on site, though not on this occasion.

3

3.20

Discussion Point

The client is concerned that the forecast final account figure is the same as the tender sum. In the light of the problems on the project, how would you advise the client?

The client's quantity surveyor may be mindful that there are some elements of the project where savings are likely to be made. For example, there may be some monies still available in the remaining groundworks, and a large part of the project comprises nominated subcontractors on which savings may have already been identified. A breakdown of the forecast final account is not given for each element (e.g. preliminaries down to nominated suppliers). It may be prudent to ask the QS for these figures.

3.21

Discussion Point

How does the contractor propose to secure this net gain? What is the consequence for the construction process of seeking such gains?

The sum of £117,317 would appear to be a substantial net gain. The majority of that is on subcontractors (£1,642,241 – £1,560,128 = £82,113) and a further £11,947 on materials (£106,947 – £95,000). It would appear that the contractor is going to make extensive use of dutch auctions, asking its suppliers for round 2 bids (bid for an order) lower than round 1 bids (bid for a bid). Our captains of industry frown upon the use of dutch auctions; they argue that it is an expensive waste of professional staff time, and the cost of this bid and rebidding process is passed on to clients. Clients who receive bids higher than they should be may be deterred from proceeding with developments: a loss to the whole economy.

3.5 Quality management

3.5.1 Introduction

Quality management is able to influence all the parameters of production for better or for worse. It may be argued that some areas of the construction industry need quality management more than most. Although certain projects, such as the London 2012 Olympics, can deliver excellence, all too often the industry is associated with poor management and a lack of control of key aspects of production: projects are often delayed, project costs overrun and spiral out of control, work is of poor quality or even defective, its health and safety record is poor and there is seen to be a lack of environmental consideration.

For example, when the Museum of Liverpool opened in 2011, visitors were unable to use the main steps to its entrance due to defective design and construction works. The client has since been awarded just under £1 million from the designer and just over £200,000 from the contractor in damages for this defective work.

Good-quality management is able to reassure the client that the project will be:

- constructed precisely as specified and designed;
- constructed in the most efficient way possible;
- delivered on time;
- delivered to budget;
- with the minimum production of waste;
- meeting all legislative requirements (e.g. health and safety and environmental legislation);
- meeting all regulatory requirements (e.g. Building Regulations and planning conditions).

There are several different levels of approach to quality management; for example, it can be implemented as a simple set of practical procedures and checklists to be used on site, or be developed into a company-wide leadership philosophy seeking to embed quality in all aspects of a company's activities.

Whatever approach is used, quality management should be at the heart of production management to make sure that clients are satisfied with their construction project from start to finish and construction companies and construction managers are working as efficiently and effectively as possible.

This section will first try to establish exactly what quality actually is, and how it can be defined and used as part of wider management systems. It will then briefly outline the common quality management standards before focusing in on how construction managers can manage quality on site. The benefits of good-quality management as a key part of production management are then set out, and how quality forms a key part of all the parameters of production.

3.5.2 What is quality?

Quality may be defined as **fitness for purpose**.

Quality may be applied to **materials**; for example, the architect will specify a type of brick for use on the project, and the bricks ordered, delivered and used in the work will have to be of this specification in terms of their quality – size, type, colour, strength. The architect will also specify a mortar and jointing technique for the construction.

Quality may be applied to **processes**; for example, combining bricks and mortar into a wall through workmanship. The bricklayers must construct the wall using the specified bricks, mortar and jointing, they should mix bricks from different pallets to ensure colour consistency, ensure that no bricks are used that are chipped or damaged, that the line, level and plumb of the wall is correct and within agreed tolerances, and that the wall is constructed in lifts (a specified number of courses of brickwork, e.g. 18 courses to a lift) to ensure that the mortar strength is achieved before more load is added to the wall structure. These details should be clearly set out in the specification for the work.

Quality may also be applied to **management** practices. In the **planning** of the work, management must ensure that the correct materials (bricks and mortar), plant (forklift to move the materials) and labour (bricklaying gang) have been procured and are available to carry out the work at the required time for the allocated cost. In the **control** of the work, construction managers should inspect the wall both during and after the construction. Scheduled inspections should ensure that the quality requirements of the materials and processes are being met during the works and, once the wall has been constructed, sign it off to confirm it has been completed to the necessary quality standards. These inspections should be recorded in order to provide documented assurance that the quality requirements of the construction have been achieved, and to enable handover of the wall to any following trades in an agreed condition.

Quality management is therefore critical in all aspects of construction management, from the ordering and checking of materials on to site, to the planning of the work, to the management of the workforce, and finally to the inspection and checking of the work once complete.

Discussion Point

How would these considerations affect other construction activities? Consider the construction of raft foundations or the fixing of roof trusses.

3.5.3 Defining quality management

Quality management has been associated with many terms and definitions:

- Quality control
- Quality assurance
- Total quality management.

These different terms refer to different levels of quality management.

Quality control

This is the most basic form of quality management, and involves the control of quality through management activities. In reality, this means carrying out inspections of work or sampling of materials, known as statistical quality control, to ensure that the requirements are being met.

Quality control aims to reduce client complaints and improve the reliability of the work being carried out, often through a focus on the detection and removal of defects in the work. Although the identification of any defects is effectively the identification of waste, as work will have to be redone to remove them, quality control should save money in the long term by eliminating the need for any future remedial works, and this does help develop client confidence as their finished products are defect free.

Quality assurance (QA)

QA builds on the ideas of quality control, but aims to provide **assurance** to the client that the quality standards for the project are constantly and consistently being met through planned quality management systems, which are clearly recorded.

QA is often undertaken through the adoption of recognised international quality management standards, the **ISO 9000 standards**. These standards provide an acknowledged framework for the implementation of quality management.

To meet ISO 9000 standards, all activities, processes and systems need to meet recognised quality requirements. They must also be recorded and documented in such a way that they can be checked to show they are being carried out as part of everyday practice. Companies may then be independently audited against their own quality systems, and be accredited and certified as a quality assured company.

There are three steps to quality assurance:

- Clearly saying what you are going to do to achieve quality.
- Writing it down as a formal quality management system.
- Proving you are actually doing it.

There have been some criticisms of QA; people think it is unnecessary and just makes more paperwork. But this can be a result of quality systems being set out and recorded by people who don't even do the job, resulting in systems that don't work in the same way that people do on sites. Setting up quality management systems for construction companies should involve the construction managers who have to carry out the work. This should result in a QA system that actually matches current work and doesn't add too much extra paperwork. If set up properly, the QA forms may be used as the site check sheets for the construction manager when carrying out inspections. Developments in IT will also help speed up the process, and inspections can be carried out on tablets, PCs and phones which can automatically update the records back in the office from out on site.

An example of a QA check sheet for a concrete pour is given in Table 3.5.1.

Total quality management (TQM)

Total quality management is a high-level leadership approach to quality management. It includes all the aspects of QA, but also tries to improve on them every time work is carried out. This notion of **continuous improvement** means that a company is always trying to improve on its quality management systems and delivery.

Continuous improvement is based on a process known as the Deming cycle. Figure 3.5.1 shows the four stages in the process: plan, do, check and act.

The Deming cycle is used to test new ideas to improve quality and performance; notice that the 'do' stage of the cycle is not putting the plan into complete action but trying it out to see whether any change is beneficial. Once real benefits can be seen, the process or procedure is then incorporated into the company management systems under the 'act' stage.

Improving quality means removing waste from the production processes, and detecting and removing defects from the work. Even better, the workforce should be trained to prevent defects from occurring in the first place and instead produce quality workmanship that is 'right first time'.

Quality may be included in all aspects of construction company activities, not just on their sites. Quality may be built into the construction process from tender preparation, to contract award, to design, to construction, and to handover and client aftercare. All of these aspects should be right first time, eliminating waste in the process, and therefore result in a positive experience for construction clients, giving them a quality construction product that meets their needs while still providing value for money.

3.5.4 Quality management standards

ISO 9000 is the internationally recognised quality management and quality assurance for all industries, not just construction, developed by the International Organisation for Standardisation.

Rather than setting down strict steps to be followed, the ISO 9000 approach allows companies to develop their own quality management processes and systems by providing guidance as to how best to achieve quality in their usual operations and to meet the basic requirements of quality management. This means that construction

Table 3.5.1 Example QA checksheet for a concrete pour

Task:	Location:
Checked by:	Date:
Print and sign:	
Foundations	**Pre-stressing**
Clean:	Anchorages:
Firm:	Ducting:
	Connections:
Joints	Vents:
Type:	Open ends protection:
Cleanliness:	Free passage:
Roughness:	Marking:
Treatment:	
Waterstop:	**Inserts**
Formwork	Civil:
Category:	Electrical:
Joints:	Mechanical:
Cleanliness:	Others (check dwgs):
Tolerance/position:	Identification:
Bracing/anchoring:	Position:
Releasing agent:	Fixing:
	Protection:
Reinforcement	**Others (list)**
Type:	
Size and spacing:	Setting out: line level and plumb:
Concrete cover:	
Tying:	
Splice laps:	
Cleanliness:	
Welds:	
Mechanical splicing:	
Comments:	

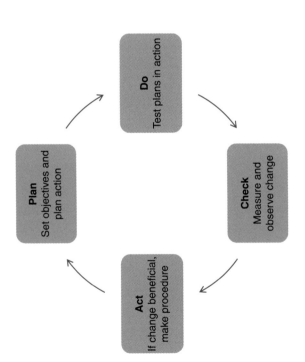

Figure 3.5.1 The Deming cycle

companies that implement ISO 9000 can develop their own quality systems, specifically for construction work.

There are different standards in the ISO 9000 series, which at the time of writing include the following:

- ISO 9000:2005 – covers the basic concepts and language.
- ISO 9001:2008 – sets out the requirements for quality management systems.
- ISO 9004:2009 – focuses on how to make quality management systems more efficient and effective.
- ISO 19011:2011 – sets out guidance on internal and external audits of quality management systems.

More details and information about the series may be found on the ISO website.

To become quality certified, companies must meet the requirements of ISO 9001. This standard involves the establishment, implementation and conformance to a **quality management system**. A quality management system should support consistent quality management in all aspects of a company's operations in order to meet client needs and other regulatory and legislative requirements. Quality management systems should also try to continually improve the processes and controls in place, and assure clients that the company is committed to quality.

Many large and medium-sized contractors and subcontractors are ISO 9001 accredited, which ensures quality assurance all the way along supply chains. QA certification can be a pre-tender requirement for some clients, who want to make sure that their projects are carried out by companies serious about quality management. QA certification also helps companies market themselves, and attracts more clients.

3.5.5 Managing quality production on site: planning and control

For the construction manager, managing quality on site forms a key part of production management.

There are a variety of processes and systems that the construction manager can put in place to ensure quality management. If the company is quality assured (ISO 9001 certified), these systems will already be in place, and clear instructions, protocols and documentation should be available for operational teams and construction managers to use. In smaller companies without formal quality systems, construction managers should establish their own set of processes to ensure that they are able to effectively manage quality on their sites.

Managing quality on site therefore involves the construction manager in the following tasks:

- Establishing the benchmark.
- Undertaking inspections and/or testing of the work.
- Recording the findings.
- Taking corrective action if necessary.

As previously noted, a well-trained workforce, who get it 'right first time', will help reduce the time and effort the construction manager has to spend on inspections of work. However, elements of planning are always needed to set the initial standards to be met and provide the necessary information to allow that to happen.

Establishing the benchmark

Quality management standards are found in a number of documents, and the construction manager should ensure that they are referring to the most recent revisions of both government regulations and project drawings and specifications to make sure they are up-to-date.

Benchmarks set the standards to enable construction work to be carried out right first time. Without clear standards to aim for, individual opinion as to what is acceptable quality comes into play, and opinions can greatly vary.

Building Regulations (Approved Documents)

These documents set out the construction standards to be used within England in any construction project. There are 14 Approved Documents within the Building Regulations, which include Part A Structural Safety, Part B Fire Safety and Part P Electrical Safety.

These relate to various construction activities and clearly set the benchmark. For example, Part A of the Building Regulations specifies the type and spacing for wall ties within a domestic cavity wall.

Recently, Approved Document 7 was produced which covers materials and workmanship, including details of how to establish the adequacy of workmanship and the fitness of materials for use in construction works.

All of the Approved Documents are free to download and may be accessed through the government's planning portal online.

British (European) Standards

British Standards set the standard requirements for both construction materials and their installation, although these standards are now prefixed 'BS EN'; as all standards are now set out by the European Union, they do contain National Annexes specific to the UK.

Construction materials may be specified through the British Standards. For example, clay masonry units (clay bricks) are specified within BS EN 771-1. This standard sets out the performance requirements for clay bricks, including their compressive strength, durability and dimensions within certain tolerances. Appearance (i.e. the colour of clay bricks) is not covered within the standard. Materials manufacturers can seek a 'CE' mark for their products to show their compliance to the standards, but this is not compulsory. Under the standards, all manufacturers must declare their products' level of compliance to the levels of quality set out within them.

The British Standard 8000 Series set out quality standards for workmanship on construction sites. These standards provide Codes of Practice for various trades, including BS 8000-3 Masonry, BS 8000-10 Plastering and Rendering, and BS 8000-8 Dry Lining and Partitioning.

Drawings and specifications

Project-specific information is provided by drawings and specifications, which should always conform to the requirements of the Building Regulations and often cross-reference back to the British Standards as necessary.

For example, the materials specified for a domestic cavity wall may state 'Clay Bricks to BS EN 771-1'.

Specifications should also include details of how the work is to be carried out, over and above any British Standards. Thus a specification may include 'workmanship generally to BS 8000-3' but also contain more information about the quality of the construction process, for example, that the brickwork should only be built to a 'lift height of 1.2m (max) above any other part of work at any time' and a 'daily lift height 1.5m (max) for any one leaf' or that the 'bond where not specified: half lap stretcher'.

Specifications will also include any deviations permissible in the quality of the construction. Such deviations are known as **tolerances**, and these accept the fact that construction is not as precise a process as, for example, building a clock. Construction materials have their own tolerances, bricks are not always precisely the same size, and construction processes are carried out by people and not machines, so it is unlikely that the bricklayer will be able to make each mortar joint precisely the same as the one before. Thus, for example, the National House-Building Council (NHBC 2011) specifies a tolerance of +/- 8mm maximum deviation on any length of wall up to 5m for fairfaced brickwork in terms of straightness on plan. This means the wall can vary from its **setting-out** line +/- 8mm along its length and still be acceptable in terms of quality. Tolerances are also provided for the thickness of the bed and alignment of the perpendicular mortar joints, the plumb of the wall (its vertical straightness) and the straightness in section.

It is important to check that the very latest revisions of drawings and specifications are being worked to on site. Any changes to the project will be included on the drawings and specifications and a new version issued, usually indicated by using a new revision letter in alphabetical order. Thus drawing 001 Rev A should be superseded (replaced) by drawing 001 Rev B. If there are a lot of drawing revisions on a project, the whole alphabet may be used up, and drawing revisions AA or BB may be used. It is also common practice not to use the letters O or I to identify revisions, as these are easily confused with zero (0) and one (1) respectively.

A document control system should form part of the site quality management system to make sure that only the latest information is used. Old drawings should be clearly marked as superseded and filed away.

Discussion Point

How would you set up a document control system on your site? What steps would you put in place?

Temporary benchmarks

New buildings must be constructed to the correct line, level and plumb, and the construction manager must ensure that these important concepts are established and maintained. At its most basic, temporary benchmarks ensure that the construction is actually in the right place!

On smaller sites, construction managers may need to be competent in the use of surveying equipment, and to be involved in levelling and setting out. On larger projects, specialist setting-out engineers are required. All setting-out equipment must be checked for accuracy before use, and that the latest drawings and specifications are being worked to.

Levels are established from ordnance **datum** levels. Ordnance Survey describes benchmarks (BMs) as:

marks made [...] to record height above Ordnance Datum [...] BMs are found on buildings or other semi-permanent features [...] Bench marks are the visible manifestation of Ordnance Datum Newlyn (ODN), which is the national height system for mainland Great Britain and forms the reference frame for heights above mean sea level. ODN is realised on the ground by a network of approximately 190 fundamental bench marks (FBMs) [...] and [...] approximately 500 000 'lower order' BMs [...] the [...] 190 FBMs

(Ordnance Survey 2014)

If a lower order BM is some distance from the site, construction managers need to ensure that a temporary benchmark or datum is established close by, probably on a permanent structure just outside the site boundary. After it has been checked and rechecked, that will be the master benchmark from which all levels on site are established.

As work progresses, a series of temporary benchmarks will be established on site, perhaps on existing manholes or parts of the new structure as it rises from the ground. On a day-to-day basis, levels for work being constructed will be established from on-site temporary benchmarks. Any errors in levels attributed to temporary benchmarks can be catastrophic, and the process of checking and rechecking these is vital. On larger projects, it is also good practice for levels or datums to be marked on walls or columns on all floors. These take the form of horizontal lines set at, say, 1.00m above finished floor level, thus allowing internal trades to easily establish the heights required for their fixings. Figure 3.5.2 illustrates a datum on steelwork. Here the column is marked as GL – Grid Line – 12A, with the level (81.810) marked as +1m above finish floor level by the horizontal line below.

Buildings must also be set to the 'building line': their correct **line** as viewed on plan.

The line for a new building may be set at a fixed distance from and parallel to an existing road kerb line. To ensure the correct line, that distance must be set exactly, and the building must be parallel and not tapered to the kerb line. Alternatively, a new building façade may be

Figure 3.5.2 Site levels marked on steelwork

set to line with the façade of one or more existing buildings; it must not protrude or be set back from that line. From the building line all individual elements (e.g. walls, kerbs, drains, rows of piles or columns) will have their own line. Lines of some elements will be at 90 degrees or some other designed angle to the main building line. Drawings may indicate grid lines, or site setting-out engineers may set a series of temporary baselines on projects; those lines may be for the external walls of buildings, or centre lines of steel columns, or centre lines of roads.

Marking for baselines may be marks on existing structures, 'hilti' nails in temporary or permanent pavements, or nails in timber pegs set in concrete within the site boundary. Also set on site are 'stations': fixed points marked with nails, on or off baselines, that have coordinates in the northeast quadrant of the compass (e.g. 10,500 metres N: 15,650 metres E). All points of buildings have their coordinates, such that angles and distances may be calculated from stations to points on buildings that need to be set out. Temporary baselines and stations must be checked and rechecked; again errors can be potentially catastrophic.

Building must also rise from the ground **plumb** or vertically; not leaning.

On single- or two-storey structures, plumb is often just assured by the bubble on the tradesman's spirit-level. For taller structures, it may be necessary to rely on the plumbing facility in EDMs.

Samples

A different kind of benchmark may be required by designers who may specify the need for samples of construction materials and processes so that they are able to set a quality standard in terms of appearance.

For example, with brickwork this will involve the construction of a sample panel on the site by the brickwork contractor from the specified bricks and mortar. This sample panel is then inspected by clients and designers and is either approved, or elements within it, such as colour mismatch, chipped bricks or poor jointing, are noted as not acceptable for the finished product in terms of quality.

Samples of other materials may also be provided as smaller versions or sections of the products to be used in the construction, such as powder-coated finishes to window frames and doors or flooring samples; others may be sampled within the construction, such as various paint samples on an internal wall of the building.

Undertaking inspections and/or testing

Inspections should be carried out at appropriate times and at key stages of the work; for example, when a work activity is completed or when one trade will take over 'ownership' of a work area from another. All subcontractors should carry out their own quality inspections before handing work over, and the construction manager's inspection should not be a replacement for subcontractors' own quality management processes.

Formal inspections may be required by Building Control who may wish to inspect and sign off key elements of the work to ensure that they comply with Building Regulations. For example, the foundations may require inspection before they are covered, and such inspections should be planned as necessary and notice given in good time to Building Control to avoid delaying the project.

Testing for various elements of the work should be carried out as specified. Often concrete is tested by the collection of samples from the delivery loads. These samples are made into cubes, which are then cured and crushed at specific intervals to ensure that the required strength has ultimately been achieved. Concrete testing may be carried out on site for large projects, or subcontracted to specialist engineering firms who undertake the tests in their laboratories.

Most inspections are visual, and in order to make sure all quality elements are included a checklist should be used as a prompt. This checklist should relate directly to the quality requirements for that work element.

Discussion Point

What sources should you use to prepare your checklist?
What would you do if the sources did not match in their requirements?

Inspections should be carried out once work is complete. For example, for the handover of completed facing brickwork the checklist used should confirm the following:

- Correct brick and mortar as specified.
- Visual comparison to the sample panel:
  - overall colour acceptable;
  - no colour bands or patches.
- Setting out checked:
  - plumb in tolerance;
  - line in tolerance.
- Perp joints plumb, equal and acceptable.
- Bed joints equal and acceptable.
- Correct positioning and inclusion of DPC.

Discussion Point

Draw up a quality inspection checklist for the window installations. Note the sources that could have established the benchmark.

Some aspects of visual inspection are a matter of opinion and cannot be measured so easily as tolerance or line, level and plumb. For example, decoration has specific quality issues such as brush marks in the paint and woodwork 'grinning through' (when the grain of the wood can be seen through the paintwork), which are less easy to quantify in terms of what is acceptable and what is not. The construction manager's experience can be useful here, or alternatively a sample room may be decorated and agreed with the client and designer to set the necessary quality standards to be measured against.

On site, quality inspections are often referred to as **snagging**. Snagging is the construction site term for the checking of any works to ensure that they meet the quality requirements. The construction manager should ensure that snagging works are minimised by good workmanship as the project progresses, but they should also produce snag lists for various trades to ensure that any quality issues are corrected as soon as possible. The project should only be handed over to the client once it is free from defects and poor-quality work.

As a result, the term 'snagging' is also used to mean the process of quality management carried out on an almost completed building by the construction manager and their team, to make sure that all the final elements of the work have been finished to the necessary standards before the handover to the client. Snag lists may be produced on a room-by-room basis, and so may involve the work of several different trades, creating a final 'to-do' list for the project. An example of such a snag list is given in Figure 3.5.3. For contractors aiming to provide

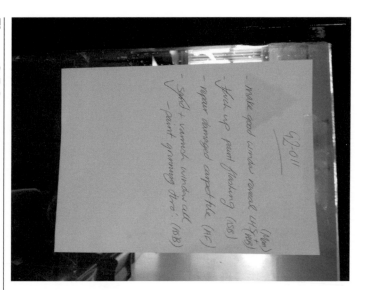

Figure 3.5.3 Example snag list

a quality product, snagging forms a vital part of their work to ensure defect-free construction projects on completion.

Clients can also produce their own snag lists and people buying new houses can also undertake their own snagging to send back to the house builder, who then undertakes any agreed works to finish the project to the new owners' satisfaction. This is a more reactive approach to quality management, and as house builders become more concerned with client satisfaction, many have moved snagging into their own stage of the works to ensure that they hand over snag-free homes.

Inspections should also be carried out as the works progress. They should be undertaken at key points within the construction works; for example, the ground-floor slab should be inspected before the concrete is poured to check that the quality requirements in terms of gas membrane, vapour barrier and reinforcement installation have been met before they are covered by the concrete. Issues of poor-quality lapping of membranes and inadequate concrete 'cover' (the depth of concrete over the reinforcement) can cause serious problems once the project has been completed. This is therefore a vital quality inspection that must be carried out before all major concrete pours. Checking that the correct ducts and drainage have been installed is also advisable to avoid costly and timely breaking out and rework later in the project. Following the inspection and quality checks, a permit to pour may be issued and the work carried out.

Inspections can also form part of the quality management of the project as a whole; for example, work areas may be inspected and signed off before the next trade moves in, creating a record of the state of that work area in case any damage is caused by the next trade. Such records make a construction manager's role much easier should remedial work have to be carried out, as this can be charged directly to the responsible subcontractor. Creating a system of sign-offs for work areas may seem like unnecessary paperwork, but it can prove invaluable in helping to control the work.

Recording the findings

Records should be kept of all agreed quality standards, all inspections carried out and all sign-offs and handovers of work areas. This can produce a lot of paperwork, but with new technology it can easily be digitised and doesn't have to result in files of paper if the records are kept on a server and backed up regularly, or even in the cloud.

Records enable the construction manager to have full control of the quality management of their project, as well as provide verification of the quality management system in place. They may also prove invaluable in tracing back the causes of any problems that may arise in the future, long after the project has been completed and the client moved in. Depending on the contract, contractors may be liable for defects for many years after the building has been finished.

Taking corrective action if necessary

For the construction manager, quality management is not just about setting and checking quality standards on their site. Control must also be employed and corrective or remedial action taken where necessary to meet the quality standards of the project.

For example, an incorrectly positioned light fitting can mean the relocation of the cabling and fitting but also the patching of the ceiling and its full redecoration to avoid the paint 'flashing' (if a small patch of new paint is applied to an already painted surface it can often be seen, known as flashing). If poor workmanship is at fault then the subcontractor concerned must bear the cost of the rework and try to recoup any lost time through the addition of resources. If the fault lies with the main contractor, for example, through issuing incorrect or out-of-date information or instructions, then it must bear the cost of the rework and any necessary acceleration and addition of resources to keep the project on programme.

Prevention is of course better than cure, and so a good construction manager will *always* have an eye on quality whenever they are on site. Stopping work and replacing poor-quality materials or correcting poor work practices can prevent quality issues from arising in the future.

If poor-quality work is not identified until it has been completed or, worse, after the project has been handed over to the client, it can have serious repercussions. It can mean the demolition and reconstruction of entire work elements, and if necessary any work elements that followed, as well as potentially causing disturbance and disruption to the client if they have already moved in.

The aim of any good construction manager should be to hand over a snag- and defect-free building, delivered through a robust quality management system. Client expectations are increasing as quality management within the construction industry improves, and company reputations are built on client satisfaction with their new construction projects.

3.5.6 Benefits of quality management

Quality management is a vital part of all production management processes, and can benefit both contractors and clients throughout projects.

Good-quality management systems will ensure that clear processes are in place for all aspects of production management, which will in turn lead to the following:

- better management of time, so fewer delays;
- better management of cost, so fewer cost overruns;
- better health and safety management, so fewer accidents and health issues;
- better environmental management, so fewer environmental incidents;
- reduction of errors and waste in construction works;
- documented evidence that construction meets the specification;

- close control and management of any changes;
- improved overall efficiency – less wasted efforts and resources on controlling the above in ad-hoc and potentially ineffective ways.

Overall, the implementation of quality management results in savings of cost and time through developing efficiencies in construction management processes and procedures, and ensuring they, and the associated construction activities, are carried out right first time.

Clients are also aware of the benefits quality management can bring to their construction projects and so quality has become as important a consideration as tender price when clients are selecting contractors to complete their projects. Good quality management is therefore not only beneficial in terms of improved company practice and efficiency, but is also a necessary aspect of company survival in the future.

3.5.7 Quality management: summary

Quality management has developed from simple quality control on site, to quality assurance of management planning and control, to total quality management throughout the company as directed by leadership teams.

Quality is embedded in production management, in the quality of the materials and how they are used in the construction work, as well as in the planning and control of quality as the works progress. For the construction manager, quality management is a key part of their everyday production management activities. They must establish the standards the work should meet, undertake regular inspections to make sure that those standards are being met, and keep records to give clients quality assurance and manage any necessary remedial works, all with the aim of delivering defect-free projects on completion.

Discussion point comments

Discussion Point

3.22

How could quality affect the materials, processes and management of other construction activities?
Consider the construction of raft foundations or the fixing of roof trusses.

For raft foundations the quality of the reinforcement and concrete will be specified by the engineer in terms of type and strength. The workmanship may be specified within BS 8000-2:1 in terms of mixing and transporting the concrete, and BS 8000-2:2 for site work within in-situ concrete. Setting out information for the foundations will be provided on the drawings to ensure that the foundations are correctly positioned on the site, and this setting out should be checked against fundamental benchmarks or GPS to ensure accuracy. Drawings may also show ducts through foundations to carry cables into the building. Setting out reinforcement on spacers may be necessary to ensure the correct concrete cover within the foundation. Management will have to ensure that the correct plant, materials and skilled labour are in place to carry out the works – for raft foundations this may include the excavation, blinding, shuttering, reinforcement and concrete pour which will involve groundworkers and excavators, joiners and timber and fixings and steel-fixers and reinforcement and tying wire. Management control should necessitate a pre-pour inspection to check the setting out, that the depth of the shuttering is sufficient and will contain the concrete safely, that necessary ducts have been included and that the reinforcement is positioned correctly before the concrete is poured. Concrete cubes and slump tests should be taken as specified to check the concrete strength and provide assurance that it will meets the specified requirements.

For roof trusses key quality information will come from the manufacturer who will design the trusses and bracing specifically for the project. Quality materials handling and storage is essential, as trusses must be offloaded and stored vertically to ensure that they keep their strength and are not weakened at the fixings. This care must be carried through to the fixing of the trusses, and ideally the trusses should be delivered and lifted into place directly from the wagon. The fixing of the trusses to the wall plate should comply with the BS 8103-3 Code of Practice for timber roofs. Other fixings will be as specified by the manufacturer. An inspection of the trusses should be made at key points to ensure that they are not racking, and that all necessary bracing has been included and correctly fixed. The roof should then be signed off before it is covered.

Discussion Point

3.23

How would you set up a document control system on your site? What steps would you put in place?

A document control system should be simple and easy for everyone to use. Often documents are issued online via a cloud document control system such as *A-Site* which is designed for the construction industry and frequently used on large projects. Using *A-Site*, the latest version of any document is easily accessed by everyone.

But on smaller sites a hard copy system may be needed. A document control register is needed to list all the drawings for the project and what revision is the latest. An example of such a register is shown in Table 3.5.2.

When a new drawing arrives on site, the register is updated with the date and new revision of the relevant drawing. This register should be kept on the wall next to the drawing rack or filing cabinet so that people can quickly see which drawing is the latest version.

Any superseded drawings should be clearly marked with a S/S across them so that they are not used in error, and filed away.

Table 3.5.2 An example document control register

Drawing control register									
		Revision							
Dwg No	Title	1/2/2014	10/2/2014	14/2/2014	21/2/2014	1/3/2014			
001	Site layout plan	A	B	C					
002	Site section	A							
003	House Type 1 layout	A	B		C	D			
004	House Type 2 layout	A	B						

Any new drawings should also be copied and issued to subcontractors working on the project. A copy should be issued to their head office and one issued to their site team at the same time, to ensure that everyone has the latest information and there is no delay as copies are sent from the office back to the site. Subcontractors should manage their own document control systems.

Discussion Point

What sources should you use to prepare your checklist? What would you do if the sources did not match in their requirements?

First, any legislative requirements should be included from the Building Regulations; for example, the required spacing of wall ties when inspecting a cavity wall. The drawings and specifications for the project should then be reviewed and key quality elements included on the checklist, which should also include the details of any British Standards referred to in the specifications.

Sometimes a specification can be written that makes reference to out-of-date or inaccurate source information. For example, a drawing may show a stair handrail at 800mm, but the Building Regulations require 900mm as a minimum (see Approved Document K Protection from Falling) so a query should be raised to check this with the design team, and in this case the Building Regulations would override the designer's wishes.

Discussion Point

Draw up a quality inspection checklist for the window installations. Note the sources that could have established the benchmark.

Window checklist:

* Frame as specified (shape, size, colour) – specification.
* Visual comparison to sample provided – sample approval.
* Setting out checked – specification:
  * plumb in tolerance;
  * line in tolerance.
* Mastic finish to perimeter correct colour – sample approval
* Fixings at correct points and secure – specification.
* Mastic finish to perimeter acceptable – specification.

The specification could also make reference to British Standards as necessary.

3.6 Health and safety management

3.6.1 Introduction: Kieron's story

Kieron Deeney worked as a steel-fixer.

On Monday, 9 August 2004, Kieron went to work on a large construction site in London's Docklands area.

He never came home.

Kieron died as a result of an accident on site. He left behind his wife Jennifer, who he had married only 13 weeks earlier. He was 25 years old.

This section will follow Kieron's story and find out what happened to him on that day.

It will find out why health and safety management should be an **inherent** part of construction management, not an **add-on** to the process of production, and why it is vital to the UK construction industry as a whole.

3.6.2 The construction industry's health and safety profile

Kieron's accident took place in 2004. In the period 2004/2005, 68 other people also died working on construction sites in the UK. There were 39 industry deaths in the period 2012/2013 (Health and Safety Executive (HSE) 2014b). Although this is a reduction in overall numbers, each incident has a story like Kieron's behind it, and construction is still considered to be high risk with regard to health and safety. Compared to other UK industries, construction does not have a very good health and safety record.

Despite steady and significant reductions in the numbers of accidents over recent years, the number of fatal accidents remains high. Although construction employs only about 5 per cent of the UK workforce, it accounts for 27 per cent of all fatal injuries and 10 per cent of all reported major injuries. Over 0.6 million working days were lost during 2012/2013 because of accidents sustained at work.

The construction industry does not do well with regard to keeping its workforce in good health either. Over 40 per cent of new occupational cancer cases reported are from those working in construction, and the industry has high rates of work-related ill health and musculoskeletal disorders (HSE 2014b).

Sadly, Kieron's accident was not a one-off, or even that unusual. These incidents occur on sites of all sizes. Although Kieron's accident occurred on a large site, managed by a large UK contractor, many sites are much smaller, operated by SMEs or sole traders. Research has shown that health and safety management can be neglected on these smaller sites, as there is more focus on getting the job done, rather than getting the job done safely. However, as Kieron's story tells us, health and safety management is not just a problem for these small sites; it affects everyone in the industry no matter the size of your site.

You can find out more about the construction industry's health and safety profile on the HSE's website, or look at the statistical report it produces and publishes each year.

Because of these problems, the construction industry, supported by the HSE, is constantly making considerable efforts to improve its health and safety record. The various measures being put in place to manage health and safety appear to be having a positive effect, although some may be more influential that others – you just can't tell from looking at the statistics. Statistics can only tell us about incidents that have occurred; they don't tell us about all the accidents that may have happened but were prevented by good management.

3.6.3 A safe industry?

Kieron's death was caused by a fall. He was working on a concrete core within the building structure. A core forms the central part of a large building, often containing the stairs, lifts and key services. This core was being constructed in in-situ reinforced concrete using a jumpform system, built independently, ahead of the external walls and floors. An example of jumpform may be seen in Figure 3.6.1.

Kieron was fixing steel reinforcement within the jumpform system before the concrete was poured to create the final structures. This method of working means that voids in the core's walls and floors, needed for doorways, lift shafts and service risers among other things, are left open as the core construction grows.

Figure 3.6.1 Jumpform construction of a lift shaft

It was through one such void that Kieron fell over 10m to his death.

Falls from heights are the most common cause of fatal injuries in the construction industry, as is shown in Figure 3.6.2.

Falls from heights are also the most common cause of major injuries within the industry; the causes of the 1913 major injuries that occurred are shown in Figure 3.6.3.

However, when the next 'level' of injury severity is examined, it may be seen that there is a change in the most common causes.

The causes of the 3133 'over-seven-day' injuries are shown in Figure 3.6.4. 'Over seven day' means that the injured person was off work for over seven days following their accident.

Discussion Point

Can you see any trends in the causes of these accidents? Is there a link to the severity of the injuries they result in?

These 'levels' of accidents are categorised due to their seriousness under the **Reporting of Injuries, Diseases and Dangerous Occurrences Regulations 1995**. These regulations, known as **RIDDOR**, make it a legal requirement for organisations to report serious workplace accidents, occupational diseases and specified dangerous occurrences (near misses) to the Health and Safety Executive (HSE) for investigation and record.

This includes fatal accidents and major injuries, such as amputations or fractures (to other than fingers, thumbs and toes) which must be reported without delay. If a person is injured and is then off work for over seven days, not including the day the accident occurred, then it must be reported to the HSE within 15 days of the accident. RIDDOR also categorises minor injuries which do not need to be reported, and near misses, some of which do need to be reported, such as the failure of lifting equipment.

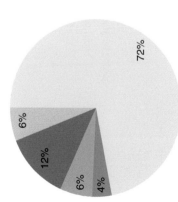

■ Falls
■ Being struck by a falling/moving object
■ A collapse/overturn
■ Being hit by a moving vehicle
■ Electricity

72%

6%

12%

6%

4%

Figure 3.6.2 Causes of fatal injuries 2012/13

Source: HSE (2014b)

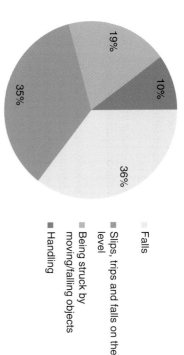

Figure 3.6.3 Causes of major injuries 2012/13

Source: HSE (2014b)

- Falls
- Slips, trips and falls on the level
- Being struck by moving/falling objects
- Handling

You can find the full RIDDOR Regulations on the government's legislation website; just search for the regulation by title.

From the statistics, it may be seen that there are many different causes of accidents on construction sites. Sadly, Kieron's accident was not unusual, and that in itself is something for the industry to be ashamed of. Although Kieron fell in 2004, in 2012/2013 most deaths in the industry were still caused by falls from heights.

Accident statistics will show us where health and safety management efforts should be focused in order to reduce fatalities – for example, in providing good edge protection to prevent falls of people and materials. Other causes result in more seven-day accidents, such as handling materials, which could be reduced by management providing more mechanical ways to move things around. In fact, there were methods in place to protect workers on Kieron's project from the risk of falls, but they were not effective. Behind each accident there will be a different series of events and actions that eventually resulted in the accident occurring. It is the role of construction managers to identify and remove or influence these events and actions before an accident occurs.

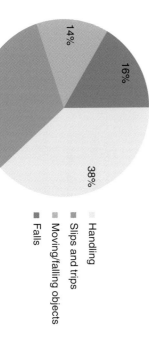

Figure 3.6.4 Causes of over-seven-day injuries 2012/13

Source: HSE (2014b)

- Handling
- Slips and trips
- Moving/falling objects
- Falls

3.6.4 A healthy industry?

Kieron's accident was a **safety** breach, and the statistics show that such accidents are unfortunately all too common within the construction industry, but the statistics also show that the industry has a poor occupational **health** record as well.

Health and safety are often grouped together as one item. People talk about 'health and safety', or 'H&S', people's jobs are titled 'health and safety manager', and this section is even called 'Health and safety management'. But they are actually two different things.

Safety is associated with accidents, like Kieron's, which are highly visible events with dramatic and devastating consequences.

Health issues are often not immediate; they may occur over long periods of time and through prolonged activity and are often not even visible; for example, lung cancers or even stress.

The high profile of accidents has often meant that safety has often been the priority in terms of 'health and safety' management and health concerns have not always been addressed in the same way, although this is changing. Considerations of health issues are growing, and health is rightly becoming just as important as safety, although it is still catching up. Many industry health issues may be directly related to the standard processes and materials used in the construction industry.

Key health issues in the industry include the following:

- Lung issues, such as diffuse pleural thickening and asthma, caused by airborne particles created by spray painting, welding, or cutting cement or brick.
- Dermatitis, either immediate (irritant contact dermatitis) or developed over time from frequent exposure (allergic contact dermatitis) to materials such as cement and acrylic sealants used in finishing works.
- Asbestosis and mesothelioma from the widespread use of asbestos in the construction industry until it was banned in the 1990s. Asbestos still remains within many buildings, and must be managed and controlled during refurbishment or demolition works.
- Cancers from silica and other materials which raise particulates when cut; although most cancers are caused by asbestos there are still concerns around other materials as well.
- Vibration White Finger (VWF), also known as Hand-Arm Vibration (HAVs), caused by the repeated or constant use of vibrating power tools, such as breakers, drills or compactors. It damages the hands and causes a loss of feeling, limited use and pain.
- Upper limb and spine/back disorder, often grouped as musculoskeletal disorders (MSDs). Caused by poor manual handing of large or awkward loads or repetitive processes, such as bricklaying, which do not necessarily involve heavy loads but do involve constant twisting or bending, as is shown in Figures 3.6.5 and 3.6.6.

Many of the symptoms of these illnesses do not occur until years after workers were exposed to the danger. It has been suggested that more cases of work-related ill health will come to light over the forthcoming

Figure 3.6.5 Bending and twisting while laying blockwork

decades. Many of these illnesses are also reportable under RIDDOR and are known as industrial diseases.

3.27

Discussion Point

Why do you think health issues in construction have not been prioritised?

Can you see any barriers to managing health in the industry?

Stress is also a health problem in the construction industry. Deadlines and coordination issues can make construction a highly stressful industry, and low reporting may be more to do with the macho culture of the sites, and a reluctance to admit to stress and well-being concerns.

The HSE (2013b) found that the top five most stressful aspects of construction work were as follows:

* having too much work to do in the time available;
* travelling or commuting;
* being responsible for the safety of others at work;
* working long hours;
* doing a dangerous job.

These elements can also be at the root of fatigue and generally poor mental health. Long hours and long commutes can also be highly

Figure 3.6.6 Bending and twisting while laying blockwork

detrimental to work–life balance, which in turn affects mental well-being. If people are not at their best, they are not efficient and effective in their work.

Such issues may also become contributory factors in other health and safety incidents. For example, although methods for protecting workers from falls were in place on Kieron's project, they were not effective. This lack of effectiveness will have been due to many different factors, which remain unknown, but if construction managers and supervisors are themselves unable to work effectively then their health and safety management will not be effective either.

The industry is starting to be more aware of these considerations and worker well-being is becoming an important issue. Some organisations are starting to offer flexible working and support for those in high-stress jobs, to ensure that they do not develop long-term health issues and jeopardise the health and safety of the working environments they are charged with managing.

There is more information about the health issues specific to the construction industry, as well as information on how to manage and prevent them, on the HSE website.

3.6.5 How is health and safety managed?

The construction industry manages health and safety in a number of ways. First, there is compliance with health and safety legislation, but the industry has also developed its own independent approaches to management, such as organisational **safety management systems** (SMSs) and **safety culture programmes**.

Approaches to the management of health and safety will vary with the size and complexity of the project; Kieron's accident happened on a very large project, which had a SMS and safety culture programme in place. On smaller projects, contractors usually just seek to comply with legislation. The use of subcontracting and long supply chains will mean that **SMEs** work on large projects regularly, and will therefore be influenced by the health and safety management practices found there. But whatever the size of the project or contractor, health and safety legislation must be complied with, if only to maintain a basic level of management and control.

Legal requirements and regulations

Health and safety management is rooted in legislation, which tries to set out what must be done to maintain a healthy and safe working environment. This legislation is not always industry specific, although there are many different regulations that apply only to the construction industry.

The overall aim of all legislation is to prevent people from being harmed at work.

Some people think health and safety legislation gets in the way of work and just causes problems, but an understanding of the legislation, rather than the myths made up by people with this negative attitude, paints a very different picture.

Legislation has been put in place to make sure that you and everyone who works with you in the industry are able to stay safe and well. People should not be injured at work: they should be able to do their job, earn their money, and return home safe and sound. What happened to Kieron should not have happened at all.

This section will introduce three key pieces of legislation:

- The Health and Safety at Work etc. Act 1974
- The Management of Health and Safety at Work Regulations 1999
- The Construction (Design and Management) Regulations 2007.

Other relevant health and safety regulations will also be noted, but cannot all be explored in detail within this section – there isn't enough space. They are summarised in Table 3.6.2, so more information may be looked up as needed.

For details and more on all legislation please refer to the HSE website or the legislation itself.

The Health and Safety at Work etc. Act 1974

This piece of legislation forms the backbone of all health and safety legislation for people at work in the UK. Under this Act, all the other health and safety regulations have been enabled, developed and subsequently implemented.

The Act was developed from a review carried out by Lord Robens in 1970 of the health and safety of people at work. Robens' report was very progressive in its approach: he suggested that negative regulation (you must not [...]) and prescriptive legislation (you must [...]) was not the best approach to modern health and safety management. Rather, he believed that risk should be managed by those who create it. He recommended that workers be involved in their own safety management, that management be committed, and that there should be a level of personal responsibility for safety.

Therefore, the Act aims to help employers decide how best to manage and control the health and safety risks in their own workplaces, as long as their actions are **reasonably practicable**. This is a legal term which means that the actions taken by employers are judged to be balanced between the risk involved and the cost (time and money) needed to mitigate it. The Act does not suggest that all risk needs to be removed; rather, any risks that remain must be managed properly.

However, some things are considered to be so important that allowing employers to decide how to manage them is not permitted – and these aspects of health and safety management are covered by various **regulations.**

The Act imposes duties upon different groups of people associated with the workplace, including employers and employees.

Section 2 of the Act states that **employers** are to ensure, as far as is reasonably practicable, the health, safety and welfare at work of all their **employees** with regard to the following:

- The provision and maintenance of safe plant and safe systems of work.
- The use, handling, storage and transport of articles and substances.
- The provision of necessary information, instruction, training and supervision.
- The maintenance of a safe place of work, with safe access and egress.
- The provision of a safe working environment with adequate welfare facilities.

Discussion Point

3.28

Can you relate each of these five requirements to common construction site activities?

Section 3 of the Act also states the general duties of employers to people who are not their employees and the self-employed, again with regard to their health and safety:

- That employers conduct their work in such a way as to ensure, so far as is reasonably practicable, that people who are not employees but may be affected are not exposed to risks to their health or safety.
- That self-employed people conduct their work in such a way as to ensure, so far as is reasonably practicable, that people who may be affected are not exposed to risks to their health or safety.

- That both employers and the self-employed must provide people with information if there is a chance that their work may affect the health and safety of others.

This section of the Act is very relevant to construction sites. Kieron was not directly employed by the main contractor who operated the site he was working on; rather, he worked for a subcontractor who was undertaking a specific part of the project works. But management have a responsibility for everyone working on their sites, including subcontractors and self-employed workers. There is also a requirement for self-employed workers to make sure that they do not put others at risk in their work as well.

In Section 7 the Act places two key duties on **employees**:

- To take reasonable care for their own health and safety, and that of others who may be affected by their actions or omissions at work.
- To cooperate with their employer and others so that they can fulfil their health and safety obligations.

The behaviour of people within the workplace is also considered within Section 8, which states that:

- No person shall intentionally or recklessly interfere with or misuse anything provided in the interests of health, safety and welfare.

These three requirements for employees mean that the Health and Safety at Work Act 1974 is not just a law designed to put all the responsibility onto employers. Rather employers and employees are equally responsible for health and safety within the workplace.

The Act also created the **Health and Safety Executive**, better known as the **HSE**, and gives powers to its inspectors, as well as setting down the offences and penalties for non-compliance. HSE inspectors can visit sites at any time and issue notices for improvements or even prohibitions, which will stop work on the site until satisfactory improvements are made. They also take those breaking health and safety laws to court. But the HSE does not try to 'police' construction sites; it tries to support industry in its own health and safety management by providing guidance, advice and information.

The HSE is government funded and has limited resources, which means that there are very few inspectors for a large number of sites; for example, in 2004 there were just 150 inspectors to police over 500,000 sites and a workforce of two million people (Smith 2004).

The HSE was notified, as was proper under RIDDOR, of Kieron's accident as soon as it occurred.

The HSE inspectors visited the site and gathered information, interviewed the people who had been working alongside Kieron on the concrete core, and tried to find out what had happened. As soon as the HSE arrived on site, they issued a prohibition notice which immediately stopped work in the core area due to the risk of falls. Nobody had witnessed the accident, but a colleague working nearby heard a loud bang. He looked through the hole in the deck and saw Kieron lying on the floor below.

Under the Health and Safety at Work Act 1974, the HSE prosecuted the main contractor in charge of the project in the London Central Criminal Court. Although Kieron died in 2004, the case was not heard until 2009.

The main contractor pleaded guilty to a breach of Section 3(1) of the Health and Safety at Work Act 1974.

Section 3(1) states that:

It shall be the duty of every employer to conduct his undertaking in such a way as to ensure, so far as is reasonably practicable, that persons not in his employment who may be affected thereby are not thereby exposed to risks to their health or safety.

The Court found that the main contractor had not taken reasonably practicable steps to make sure that Kieron was not exposed to risk. They were found guilty of a breach of the Act.

They were fined £135,000 with costs of £18,313.10 (HSE 2009).

Management of Health and Safety at Work Regulations 1999

In its reports of Kieron's accident, and the successful prosecution, the HSE stated that 'the risks of working at height and the need to manage voids in platforms are well known' and that 'risk assessments and method statements had been carried out' on Kieron's site (HSE 2009). This indicates that something else went wrong, but in order to follow Kieron's story, an understanding of risk assessments and method statements is needed to see what management practices *should* have been in place.

Risk assessments are a key way in which health and safety is managed on site, and they are a legislative requirement under the Management of Health and Safety at Work Regulations 1999.

These Regulations require employers to undertake a 'suitable and sufficient' assessment of health and safety risks of work to employees and others who may be affected by the works, such as visitors and members of the public. The objective is to determine what measures need to be put in place to remove or reduce the potential risks. Construction managers will regularly need to carry out risk assessments for elements of work under their control, as well as more fundamental site management tasks such as exchanging skips or emptying wheelie bins.

Where there are five or more employees, the findings of risk assessments must be recorded in writing and regularly reviewed in line with work activities, and amended where necessary.

This process of risk assessment (RA) forms the basis of many other health and safety legislation and regulations, and has become the standard management tool for risk reduction and the establishment of **safe systems of work.**

Process of risk assessment

The HSE (2012b) suggests 'Five Steps to Risk Assessment':

1   Look for the hazards.
2   Decide who might be harmed, and how.
3   Evaluate the risks and decide whether existing precautions are adequate, or whether more should be done.
4   Record the significant findings.
5   Review and revise the risk assessment as necessary.

Look for the hazards

There are two new definitions here: a **hazard** is the potential for something to cause harm, whereas the **risk** is the likelihood of that something actually causing harm.

For example, the edge of a slab is a high hazard – should a fall occur, there is the potential for serious harm. However, with edge protection in place it becomes low risk – people cannot get to the edge to become victims of the hazard. An example of this is illustrated in Figure 3.6.7.

These two terms are often confused, but although hazards will often always exist, such as the edge of a slab, the way the risk of the hazard is managed is what will keep people safe.

Decide who may be harmed, and how

The risks may just be to the operatives carrying out the work, but others may also be affected. Many different trades with many different risk assessments may be working on the same site and may be

Figure 3.6.7 A protected slab edge

affected by each other's works. Members of the public may also be at risk.

Some people also need special consideration with regard to risks, such as young people (under 18 years of age), expectant mothers, people with disabilities and people working on their own, known as lone working. In these instances specific risk assessments must be carried out for these individuals and their work as required by the Regulations.

Risks may not just be associated with accidents and safety, although this must of course be considered. A risk assessment may also reveal potential health concerns with specific work tasks. For example, long-term use of power tools could create a risk for operators suffering from Vibration White Finger. In this case, maximum use times must be stated in the risk assessment for the operatives affected.

Evaluate the risks and decide whether existing precautions are adequate, or whether more should be done

Risks can be evaluated with regard to their likelihood of occurrence and prioritised as high, medium or low risk. Different levels of risk relating to one work activity can often be reduced together, by following a **hierarchy of risk control**.

The hierarchy of risk control considers how risks may be reduced through managing work practices. It uses the acronym ESCAPE, as is shown in Figure 3.6.8.

Risks can be assessed following the hierarchy of control; number 1 is the most effective method – to eliminate the risk altogether. Personal protective equipment (PPE) as a management tool is only number 5 on the hierarchy. If there is no way to reduce the risk, then educating the workforce as to the risks and enforcing the devised safe system of work becomes essential.

Discussion Point

Consider the following construction activities; how can the risks be reduced through application of the hierarchy of control?

1. The movement of a forklift truck around a site.
2. Fitting edge protection to a steel beam on the edge of a first-floor slab.
3. Spraying intumescent paint on exposed soffit steelwork.

Record the significant findings

Many large organisations have a standard risk assessment form which is used for all work activities. The HSE also has examples on its website.

An example risk assessment for a manhole excavation is shown in Table 3.6.1.

Table 3.6.1 Example risk assessment for a manhole excavation

PROJECT:	New Build School						
TASK:	Manhole Excavation						
	Hazards	Who Might be harmed and how?	Evaluation		Precautions to be Taken	Action By	Residual Risk
			Hazard	Risk			
1.	Interaction with Machine (excavator and dumper)	All site operatives Struck-by injuries	H	M	Machine to be banked at all times. Work area to be actively segregated by barriers to avoid other operatives accessing the area.	Fred Bert	L
2.	Collapse of Excavation	Site operatives working on task Crush injuries	H	M	Excavation to be supported using a manhole box, securely positioned, before any personnel access the excavation.	All	L
3.	Fall from height	All site operatives Injuries from the fall – breaks or twists etc.	H	H	Only briefed operatives to access work area adjacent to excavation. Work area to be actively segregated by barriers to avoid other operatives accessing the area.	All Bert	L
Assessed By:	Fred Sherratt	Date	13/02/13		Approved By:	Peter Farrell	Date	15/02/13

Note: This RA is to be reviewed if there are any changes of work method, changes to the work environment or deviation from the method statement.

Review 1:		Date			Review 2:		Date	

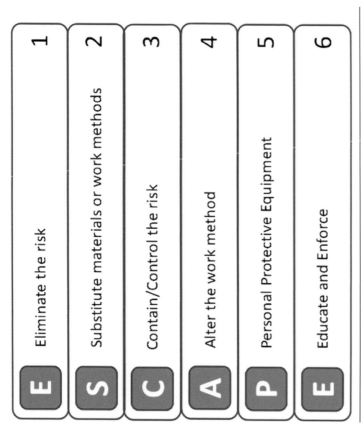

E	Eliminate the risk	1
S	Substitute materials or work methods	2
C	Contain/Control the risk	3
A	Alter the work method	4
P	Personal Protective Equipment	5
E	Educate and Enforce	6

Figure 3.6.8 ESCAPE hierarchy of risk control

Review and revise the risk assessment as necessary

Risk assessments must be reviewed when the work environment or task changes from that originally foreseen. Construction sites are by nature places where things change quickly.

Sometimes risk assessments are just seen as a paperwork exercise; people simply use the same one for all their work on all the sites they visit, or they cut and paste the content on to a new form. This renders risk assessments useless.

Risk assessments are vital in ensuring that all aspects of work have been considered and made as safe as possible. The process of carrying out a proper risk assessment often makes people think carefully about their work, and may highlight new issues and potential risks – it is the process rather than the paperwork that is important.

In many cases, risk assessments show that a safe system of work has been decided and put into place.

Safe systems of work are also effectively demonstrated through method statements (MSs). Main contractors will often request the submission of subcontractors' risk assessments alongside a method statement for their works on the site. Together, the two are sometimes referred to as **RAMS**.

A method statement describes the work to be undertaken step by step, and the notable risks which emerge from this process can then be highlighted and recorded on the associated risk assessment. It should also include company information, key contacts and their phone numbers, the overall scope of work, equipment and materials to be used and the necessary operative personal protective equipment (PPE). Although method statements are not legally required under any regulations, they are useful in demonstrating compliance and clearly communicating work processes.

Exercise Q

What hazards and risks can you see in Figure 3.6.9?

How do you think they are being managed, and what would you do differently?

Undertake a risk assessment for the works you see in Figure 3.6.9 following the HSE's five-point plan.

Suggested solutions to the exercises may be found in the Appendix.

Communication of the RAMS

Although the paperwork involved in RAMS helps construction managers reduce health and safety risks on their sites, this information must also be passed on to those carrying out the work.

RAMS briefings should be held before operatives start any new task or begin work on the site. Construction managers or subcontractors' supervisors or foremen must ensure that the RAMS have been read and understood so that the operatives are able to recognise why they have to follow a particular system of work in order to remain safe and healthy.

Figure 3.6.9 Scenario for Exercise Q

An even better approach is to get the operatives involved in the production of the RAMS. They are going to be carrying out the work and will likely have valuable experiences to bring to the process.

The HSE found that RAMS had been produced on the site where Kieron had his accident (HSE 2009). The hazard that had been identified was a void within the core. The risk that had been identified was that someone could fall through the void to the basement floor of the core below.

Such risks can be eliminated by covering the void with a fixed cover or the use of edge protection, and, according to the Risk Assessment findings, weekly and monthly inspections should have been carried out. So what went wrong?

Although the RAMS will identify the hazards, the risks, and the way health and safety management will remove them both, this process must also be **communicated** and **acted upon** for it to be effective. Without communication and action, RAMS become a paper exercise, and useless.

The HSE found that the monthly and weekly checks required in the risk assessment were not being adequately carried out (HSE 2009). There was also insufficient edge protection within the jumpform to prevent falls, and the area was not kept free from hazards.

An unknown person had covered the void with a piece of poor-quality plywood. This did not provide enough protection for the void, but it had not been spotted in any of the checks.

It was through this void that Kieron fell.

The Construction (Design and Management) Regulations 2007

Another key piece of construction legislation that affected the site where Kieron had his accident is the Construction (Design and Management) Regulations 2007, often abbreviated to CDM2007.

These Regulations focus on both the design and management aspects of construction. They place health and safety duties on all parties involved in a construction project, including the client, designer and contractor, and creates a CDM coordinator to ensure that health and safety is considered and managed at all stages of the project from conception to operation.

For example, designers must consider health and safety as they are designing their elements of the project. They need to carry out risk assessments about how their designs will be built, and let those constructing the project know what health and safety risks remain. It has been suggested that 50 per cent of accidents could be avoided by a change made in the design of a project.

More information about these duties may be found in the HSE's Approved Code of Practice for CDM2007; this may be downloaded for free from the HSE website.

CDM2007 also places certain duties on contractors and construction managers, who must plan, manage and monitor their works so that they are, so far as is reasonably practicable, carried out without risk to health and safety.

Under CDM2007, some projects are termed **notifiable** where the work is likely to involve more than 30 days or over 500 person days. Many large projects fall into this category.

The project on which Kieron had his accident was a notifiable project. The CDM2007 Regulations require all notifiable projects to have a named **principal contractor** appointed by the client. Principal contractors take charge of the site, and also undertake specified additional duties. They must:

- ensure that adequate welfare is provided;
- draw up and implement formal site rules;
- ensure that all workers have a site induction;
- ensure that workers are competent and have the training they require;
- ensure that subcontractors are given reasonable time to prepare for their work;
- provide subcontractors with site health and safety information;
- consult with the workforce about health and safety matters or concerns;
- ensure that the site is secure and unauthorised people cannot access it.

The principal contractor also has to prepare the **construction phase health and safety plan.**

This plan demonstrates how the works will be managed and monitored. Like a risk assessment, the plan is likely to need revision as the project progresses. The plan should also contain the key health and safety risks of the project works, such as the prevention of falls, control of lifting operations, working in excavations, and how they will be mitigated.

The plan on Kieron's project should have contained information about the hazards found within core construction, and how these risks would be mitigated. This information would have been articulated through a risk assessment and method statement, as discussed in the previous section.

A health and safety construction phase plan is outlined in Figure 3.6.10.

Other health and safety regulations

Many regulations regarding specific activities have also been brought into law under the Health and Safety at Work Act. Some are relevant to all workplaces, and some to just the construction industry. Some regulations are necessarily prescriptive (e.g. in the management and treatments of asbestos), but others again allow management to decide how to achieve the goals set out by the regulations.

Table 3.6.2 lists many of the most relevant regulations to the construction industry, and gives examples of when they may come into effect.

Kieron's accident could also have been considered from the perspective of the Work at Height Regulations 2005. Working within

CDM Construction Limited

Health and Safety Plan
for
NewBuild School, Anystreet, Anytown AB1 2CD

Contents

1. **Description of the project**
    - 1.1.1 Project description and programme

    1.2 The project team
    - 1.2.1 Client details and contacts
    - 1.2.2 Principal contractor details and contacts
    - 1.2.3 CDM coordinator details and contacts
    - 1.2.4 Designers
    - 1.2.5 Key contractors
    - 1.2.6 Other consultants

    1.3 Existing information
    - 1.3.1 Pre-tender information
    - 1.3.2 Key issues that affect the works
    - 1.3.3 Security
    - 1.3.4 Welfare

    1.4 Management of the project
    - 1.4.1 Organisation chart
    - 1.4.2 Health and safety appointments
    - 1.4.3 Supervision standards

2. **Health and safety management**

    2.1 Management processes
    - 2.1.1 Inspections and audits
    - 2.1.2 Compliance monitoring

    2.2 Communication processes
    - 2.2.1 Inductions
    - 2.2.2 Site rules
    - 2.2.3 Minimum training requirements

    2.3 Emergency procedures

    2.4 First aid

    2.5 Incident and accident reporting

    2.6 Site risk management
    - 2.6.1 Method statements
    - 2.6.2 Risk assessment and method statement approval process
    - 2.6.3 Site-wide risk assessment
    - 2.6.4 Task-specific risk assessment

    2.7 Health and safety file
    - 2.7.1 Management process
    - 2.7.2 Guideline information for inclusion

    2.8 Key documents
    - 2.8.1 Risk assessments
    - 2.8.2 Design residual risk register
    - 2.8.3 Site layout
    - 2.8.4 Traffic management plan
    - 2.8.5 Emergency escape plan

3. **Revision register**

Figure 3.6.10 Example contents of a CDM health and safety construction phase plan

Table 3.6.2 A Quick Guide to Common Construction Industry Regulations

Regulations	Year	Affected work activities	Key points to note
[Control of] Asbestos at Work Regulations	2002	Works in existing buildings where asbestos may be present.	♦ Exposure to asbestos must be prevented. ♦ Employers or clients must find out if asbestos is present, assess work to be carried out, develop a work plan and inform the relevant authorities before work commences. ♦ Work around asbestos is highly specialised and should be carried out by those licenced to do so.
Confined Spaces Regulations	1997	Work in confined spaces, such as manholes or excavations, which are substantially enclosed.	♦ Risk Assessments must be carried out to determine a safe system of work. ♦ Permits to work are required. ♦ Rescue procedures must be established before work commences.
Construction (Head Protection) Regulations	1989	All building and construction work.	♦ Head protection is required on all sites. ♦ Employers must provide suitable head protection.
Control of Substances Hazardous to Health Regulations (COSHH)	2002	Dealing with substances (chemicals or other materials) that can be hazardous to health. COSHH substances include solvents, silicones and mastics, concrete and cement. Dust and fumes are also covered by COSHH.	♦ Risk Assessments must be carried out. ♦ This may result in exposure limits or the use of specific PPE. ♦ Control measures must be put in place. ♦ A programme of health surveillance must be provided for people who are regularly exposed to COSHH regulated materials.
Electricity at Work Regulations	1989	Working with temporary electrics and electric installations on site.	♦ The need for 'duty holders' to be responsible for electrical systems. ♦ Only competent people may work with electricity or supervise others' work.
Health and Safety (First Aid) Regulations	1981	Provision of first aid on site.	♦ Adequate provision for first aid must be made. ♦ There must always be at least one first aider on site during working hours.

Lifting Operations and Lifting Equipment Regulations (LOLER)	1998	Any lifting operations. This includes cranes and hoists, mobile elevated working platforms (MEWPS) and also excavators where they are used for lifting on site.	♦ Lifting operations must be planned by a competent person, including a Risk Assessment. ♦ Lifting equipment must be marked to show safe working loads.
Manual Handling Operations Regulations	1992	Any work involving the pushing, pulling, lifting, lowering or carrying of materials or equipment.	♦ Risk Assessments must be carried out. ♦ The working environment and individual capability should also be considered. ♦ There is no maximum or minimum weight which can be manually handled.
Control of Noise at Work Regulations	2005	Any work in a noisy environment.Many construction site activities are affected by these regulations.	♦ Risk Assessments must be carried out. ♦ Sets exposure limit values and action values. ♦ Hearing protection must be provided by employers.
Personal Protective Equipment at Work Regulations	1992	The use of any PPE at work	♦ If assessed for, employers must provide PPE. ♦ PPE must be suitable, maintained and inspected on a regular basis. ♦ PPE must be provided along with information, instruction and training for its use.
Provision and Use of Work Equipment Regulations (PUWER)	1998	Where any work equipment is used.PUWER also applies to lifting equipment, in addition to the Lifting Operations and Lifting Equipment Regulations (LOLER).	♦ Equipment must be suitable for the task. ♦ Equipment must be regularly inspected and maintained. ♦ Information, instruction and training must be provided.
Reporting of Injuries, Diseases and Dangerous Occurrences Regulations	1995	When an accident or incident occurs	♦ Refer to Section 1.3 for details of this regulation.
Work at Height Regulations	2005	Any work at height	♦ Risk Assessments must be carried out. ♦ Suitable equipment must be selected. ♦ Falls of objects from height must also be considered.

the core would mean working at a height above the ground, and in an area where there was the risk of a fall. Under the Work at Height Regulations, a risk assessment must be carried out for the work, and this had been done. As previously noted, it was the ineffectiveness of this risk assessment that was the problem.

Table 3.6.2 is not exhaustive, and for more information on health and safety management in terms of the legal requirements and regulations, visit the HSE's construction homepage on its website. The contents and details of all UK legislation may be found on the government's website. Guidance and Approved Codes of Practice (ACoPs) to help follow the legislation are also provided by the HSE.

Training and certification

Many health and safety regulations refer to 'competent persons', but competence can be hard to define. One definition suggests that competence relates to knowledge, ability, training and education (KATE). While knowledge and ability often come from experience, training and education can be specifically undertaken, and will contribute to knowledge and ability in turn.

Training is a vital part of health and safety management, and ensures that managers, supervisors and operatives have the necessary skills, knowledge and understanding to carry out their health and safety responsibilities correctly.

Several industry standards have emerged in terms of health and safety training certification, and these are often supported by an organisation's own health and safety training programmes.

The Construction Skills Certification Scheme (CSCS) cards are often a requirement on large sites, and are recommended for all workers under CDM2007. To obtain such a card, workers have to demonstrate a level of knowledge about health and safety by taking an online test. The cards are linked to work experience and training for trades, or the supervisory or management level the individual has achieved. You can find out more about the scheme on its official website.

The Site Safety Plus Scheme is delivered and accredited by CITB ConstructionSkills. This includes the Site Managers' Safety Training Scheme (SMSTS) which is a programme covering various aspects of construction site management and can also be a requirement for all managers working on large sites. The course includes legislation, site establishment, risk assessments, excavations and working at height among its topic areas. There is also a scheme for site supervisors: the Site Supervisors' Safety Training Scheme (SSSTS). More information may be found on the CITB ConstructionSkills website.

On site, informal training is often carried out in the form of tool-box talks (**TBT**s). These are brief safety training sessions, so called because they are often held round the 'tool-box', and delivered by the construction manager or trade supervisors.

The subject of the TBT should be relevant to the current or upcoming tasks for the workforce, such as the use of a particular piece of equipment or PPE, or on wider site safety and health issues, such as dust inhalation or manual handling. A relevant TBT for those working in the area alongside Kieron could have covered void and edge protection.

As with risk assessments and inductions, the delivery of TBTs is very important. The construction manager must make sure that people are paying attention and the message is being communicated clearly, since both are vital to getting the information across.

Discussion Point

3.30

What TBTs may be relevant for a gang about to start excavation works?

How would you make sure that the gang is paying attention to your TBT?

Site inductions

Delivery of site specific information is given through **site inductions**, which are also a legal requirement under the CDM2007 for notifiable projects.

Inductions are the first point of contact for people arriving on sites, so they must be effective and pass on all the relevant information to minimise the risk to these new workers. The aim of the induction is to give new operatives all the information they need about the safety requirements of the site before they start working there; many accidents involve people who have only been working on the site for a few days.

Site inductions should include as a minimum:

- Basic site information:
  - location of the site;
  - site layout;
  - the management team;
  - access points;
  - traffic management;
  - site rules.

- Current work activities:
  - what works are going on at present? (e.g. steel erection or groundworks);
  - site-specific hazards;
  - access restrictions;
  - key activities (e.g. cranes on site);
  - hearing protection zones.

- Emergency procedures:
  - emergency muster points;
  - evacuation routes;
  - accident reporting.

- Training requirements:
  - CSCS cards;
  - method statements;
  - risk assessments for work.
- Environmental procedures:
  - waste management;
  - environmental policy.
- Methods of communication:
  - safety committees;
  - safety noticeboards;
  - near-miss reporting.

On small projects, inductions may just be given verbally. On some sites, booklets or leaflets containing key site information are often provided to attendees, so they have a record of what they have been told.

On larger projects, a PowerPoint presentation may be used to deliver the induction, as shown in Figure 3.6.11, and sometimes DVDs are made specifically for the project to be shown to the inductees. On very large sites, the induction can last a full day, and result in the award of a safety passport for that particular project.

Whichever method is chosen, it must be effective and those attending the induction need to pay attention to the information being provided. Operatives whose work means they are on many sites for short durations can attend a lot of site inductions, so it is vital that the induction is engaging and delivered with interest to pass on the relevant information. The most important information is that specific to the site – most sites require workers to wear a hard hat and boots, but this site may also require ear protection in certain areas owing to noisy work activities. Unless this is clearly explained, operatives are unlikely to be aware of it and may be put at risk of damage to their hearing.

The key hazards and risks for sites should be identified. On Kieron's site, the fact that the core works were ongoing would have been a key work activity, and the associated risks of falls from heights identified.

3.31

Discussion Point

What do you think are the most important aspects of the site induction?

How would you deliver your site induction to operatives?

Site inductions are seen as vital to the success of health and safety on sites. They are the best way to set out the approach to health and safety management from the first time an operative steps on to the project. Construction managers often deliver the induction personally to make sure that everyone pays attention, is aware of the rules and standards in place, and commits to working safely on their sites.

Figure 3.6.11 Induction slides

Personal protective equipment

Personal protective equipment (PPE) is one of the most common ways health and safety is 'managed' in our industry. You can tell construction workers by their hard hats and high-viz vests, even when they are not on site.

Hard hats are required by the **Construction (Head Protection) Regulations 1989** which make it a legal requirement for employers to provide suitable head protection for those working in 'building operations', and to make sure that people wear them. It is also usual for sites to insist on steel toe-cap boots, or boots with a cap and mid-sole protection, and high-visibility jackets or vests. Many large sites now also include light eye protection and trade-specific safety gloves as a minimum requirement. Any other PPE will be as specified within the risk assessment for the specific task to be undertaken, and must legally be provided by employers.

PPE is not the ideal solution to a risk, as it only protects the person wearing it, not others who may enter the work area, and relies on people wearing it correctly at all times. This is why it is the penultimate approach on the risk reduction ESCAPE hierarchy of risk control. It must also be replaced when it is no longer effective.

Discussion Point

Wearing PPE is often a site rule.
How would you encourage people to wear the correct PPE?

Some workers do not like to wear PPE, as they find it uncomfortable and consider it unnecessary, but PPE is used for a reason – to protect them from hazards – and so should be worn as required. Education and training are therefore as important a part of health and safety management with reference to PPE as the PPE is itself.

Safety management systems

As well as complying with the legislative requirements for health and safety management, many larger companies have established specialised safety management systems (SMSs). SMSs provide a systematic structure for the legislative and other safety management requirements of the company. This information is often delivered through an intranet, and is supported by training programmes to ensure that everyone knows the procedures and processes to follow in managing health and safety.

SMSs start at the highest level of the company, and contribute to the corporate management procedures. In practice this may mean including health and safety in business planning at all levels, establishing a way of measuring organisational health and safety performance, the setting of organisational objectives for improvements, and how the entire process will be audited and reviewed to enable continuous improvement and development.

At the site level, SMSs can provide structured forms for the assessment of risks and hazards, organisation site rules, induction templates, permit to work systems (where some tasks require a permit from management, as is shown by the sign in Figure 3.6.12), and communication processes such as the establishment of site safety noticeboards and regular safety committee meetings.

SMSs can also contain timetabled activities for health and safety management processes; for example, recording workforce training to ensure that health and safety knowledge is up-to-date, or scheduling and monitoring health surveillance for at-risk workers.

Regular **health and safety inspections** often form part of SMSs, and the construction manager should inspect their site on a weekly basis to check on issues such as housekeeping, access, scaffolding, public

Figure 3.6.12 Permit to work sign for roof access

interfaces, PPE, access to heights and materials storage, and take action if things are not up to standard. Inspections also enable any new hazards to be identified, recorded and acted upon quickly. These inspections are very similar to those that should have been carried out under the risk assessment for the core works on Kieron's site. A regular and effective weekly inspection should have identified poor void protection and replaced it with a proper effective covering or edge protection.

The HSE have produced guidance for SMSs in the form of their *Successful Health and Safety Management* document published in 2006. The HSE's Approved Code of Practice for the CDM2007 Regulations also provides information that can help develop SMSs.

Safety culture

The most recent development in terms of health and safety management is the aim to establish a positive safety culture on sites.

Discussion Point

How would you describe construction industry culture?
How would you describe the safety culture on sites?

Safety culture is not a new idea; it is used in many other industries worldwide, and the HSE actively encourages the development of a proactive safety culture on construction sites, seeing it as essential to help improve the health and safety record of the industry.

One definition of culture is 'the way we do things round here', so a positive safety culture means 'we do things safely round here'. In a positive safety culture, the health and safety risks of a task become its most important aspect, rather than production or working to deadlines. The health and safety of the workforce becomes the most important thing on site. Nothing is worth having an accident for.

While a positive safety culture may be seen as a good thing for everyone working in the industry, from a construction management perspective it is hard to establish whether or not one is in place. Research has shown that top management commitment and the elements considered under a SMS have a contributory effect, but it is the reality of what happens on the site which really shows whether or not there is a positive safety culture in place.

Many organisations develop safety programmes to try to develop their safety culture, such as Zero Harm, Balfour Beatty's safety programme. Balfour Beatty is the largest construction contractor in the UK, with a turnover of over £10 billion globally per year. The Zero Harm programme aims to cause zero deaths, zero debilitating injuries and zero long-term harm to workers' health.

Such safety culture programmes try to make safety personal, asking people to take part in safety management through their actions on site and choosing to work safely, rather than being forced to do so by site management.

Safety committee meetings and worker engagement processes through events or feedback cards are common practice, and seek to engage the workforce in participating in managing their own health and safety and that of their colleagues. Sites also try to develop a 'team effort' approach to safety, displaying a pride in their good safety record, as is seen by the display in Figure 3.6.13.

Developing a positive safety culture is the current focus for large construction companies, with many looking to achieve zero fatal or major accidents on their sites. Key to this culture are the company's construction managers, who must pass on the message that production and achieving deadlines are not important if someone is hurt in the process.

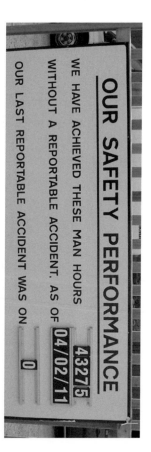

Figure 3.6.13 Safety performance displayed outside a site

Safety culture is something that Jennifer Deeney, Kieron's wife, has talked about a lot since his death. In her time visiting sites following Kieron's accident to talk about safety, as part of an industry safety culture programme, Jennifer found that one of the main causes of accidents is assuming that someone else will take responsibility for safety. Health and safety should be *everyone's* responsibility. People should not assume that someone else will make something safe. People should not walk past something like the poor-quality piece of plywood covering the void through which Kieron fell. People should speak up and take action. This is at the heart of safety culture – the idea that everyone needs to be involved and work together to keep each other safe.

3.6.6 What happened to Kieron?

Kieron went to work on Monday, 9 August 2004. He never came home.

In following Kieron's story, it may be seen that health and safety legislation had tried to put in place measures to prevent an accident like this from occurring.

The main contractor was under an obligation to ensure that people affected by their work were not exposed to **risks** to their health or safety under the **Health and Safety at Work etc. Act 1974**. A **risk assessment** and **method statement** had been carried out for the area in which Kieron was working, as required by the **Management of Health and Safety at Work Regulations 1999**. This document identified that there were safety **hazards** and risks in the area, and proposed methods to remove and mitigate them. Weekly and monthly checks had been scheduled, an inherent part of a site-wide **safety management system**, to ensure that the edge and void protection was in good order.

But this is where the system broke down.

These checks were not carried out adequately. The system to make sure that there was sufficient edge and void protection failed.

Someone had tried to resolve the problem. A poor-quality piece of plywood had been placed across the hole. It was not suitable void protection, it wasn't fixed and it did not have enough bearing around the edges. No one knows who placed the plywood there, or why.

Somehow, Kieron fell through the plywood and the void, over 10m to the basement floor below.

Why do accidents still happen?

Kieron's story demonstrates that there is more to health and safety than just legislation. The law is there to try to stop accidents and poor health in the workforce.

But there are other factors involved, and a lot of research has been conducted in trying to explore what these factors are and how best to manage them.

Even with the best intentions in terms of risk assessments and method statements, **human factors** still come into play. People make mistakes. Things can never be 100 per cent foolproof.

On construction sites, there is always the pressure for constant production. Time and cost are two key aspects on all construction projects, and there are pressures on both management and the workforce to ensure that projects are completed on time and on budget.

This makes construction sites vulnerable. The wrong things can be prioritised. The need for speed can make people cut corners and take risks. The complexities and pressures of managing the work can mean that the focus on production takes the attention away from performing safety inspections. Cost cutting may indicate that the right piece of access equipment is too expensive, so that people have to 'make do' with what is available on site. Work may be paid on 'price' and operatives may be reluctant to stop work for safety reasons if it means that they will lose money. The correct PPE may be too costly. There may not be enough managers and supervisors on site to manage the health and safety properly, in addition to all their other work.

A delay or overspend can cost organisations time and money. But a delay that costs a life should be far more serious.

The cost of an accident can be measured in financial terms or in people terms, but either way safe and healthy sites will attain good productivity as people can work to their best in a safe and healthy environment.

Having the wrong priorities is just not worth it.

How can we stop accidents on site?

As well as following legislation, keeping training up to date, and using other best practice processes such as SMSs on sites, construction managers must also manage people.

From Kieron's story it may be seen that the processes were in place to make the area Kieron was working in safe. It was people who caused the break in the system.

Construction managers must help to create sites where the processes and practices of health and safety management can be carried out effectively. That may mean they must challenge others to obtain the correct piece of safety equipment, and stop unsafe working no matter how fast the deadline is approaching. They may have to demand more time or support from their head offices so that health and safety management can be carried out correctly.

Most importantly, they have to communicate the health and safety information effectively to the workforce.

They must pass the message on that health and safety is the **number one priority** on their site.

They need to motivate and encourage people to participate in keeping themselves and their colleagues safe and well.

They must '**walk the talk**' and put their own views about health and safety management into practice. They must be brave enough to speak up and stop work for health and safety reasons. They need to manage the site to develop a positive health and safety culture for everyone's benefit.

3.6.7 Health and safety management: summary

This section has looked at the many ways in which health and safety is managed within the construction industry. Legislation and regulations play a large part in setting out what is required in terms of minimum standards, and in some cases more detailed protocols.

The competence required by many regulations has resulted in the establishment of many industry training standards, such as CSCS cards. In addition, many larger organisations have also developed their own safety management systems and are seeking a positive safety culture through their own training programmes and on-site schemes.

In practice, health and safety forms a vital part of construction management. Management skills are required in the identification of hazards and risks, and in the clear and effective communication of risks to the workforce. Construction managers also need to ensure that strong safety cultures are developed on their sites, where health and safety is everyone's problem. Everyone needs to work together, and not to walk on by, thinking that someone else will sort out a problem. Someone else may not, and another unnecessary death like Kieron's could occur.

This section has been produced with the support of Kieron's wife, Jennifer Deeney. Jennifer had only been married to Kieron for 13 weeks. Fourteen weeks after her wedding day she was standing in the same church to see him buried. Jennifer speaks out about Kieron because in the UK construction industry accidents are no longer news. They don't make the headlines because they occur on an all-too-regular basis. She believes that people should be made aware that they need to look after each other on site; people need to think about their families and friends and not walk on by.

Health and safety management is a **vital** and **inherent** aspect of construction management.

Anything the construction manager plans or puts into practice should consider the potential health and safety repercussions and ensure that any risks are managed correctly.

Health and safety management should be something construction managers **just do.**

Anything else is just not worth it.

Discussion point comments

The types of incident appear to have an effect on the resulting injury. Falls from heights tend to have more serious consequence and are the major cause of both fatalities and major injuries. Falls from height also cause seven-day injuries, but they are not the biggest cause. Poor handling is the most common cause of seven-day injuries.

The top four causes of accident are common to major and seven-day injuries, but in different proportions, showing clear hazards and areas for management focus. Being struck by a moving/falling object is also common to fatal accident causes, further emphasising the need for management of working at height. There are some unique causes for fatal accidents, suggesting that being hit by a vehicle or a collapse is likely to be a serious event, resulting in death.

Discussion Point

Can you see any trends in the causes of these accidents?
Is there a link to the severity of the injuries they result in?

Discussion Point

Why do you think health issues in construction have not been prioritised?
Can you see any barriers to managing health in the industry?

Health issues are long term and not immediate events. Safety is very easy to see, and very easy to see where it has gone wrong. Health issues often take a long time to become evident and in some cases the link between poor health and work can take some time to be made, for example, in the case of asbestos which was used widely in the industry until associated health problems began to appear years later.

Health issues may not have been prioritised as they can be harder to manage, and people were less aware of the activities that caused them. The macho culture of the industry may also mean that people are reluctant to admit they are ill or have a health issue – they may think this means admitting that they are not 'tough enough' for the job.

This can also cause problems in managing health on site. People often don't see the long-term problems that might happen – if they are well today then that's all right, rather like smoking. People can also be reluctant to wear protective gloves if they have always used their bare hands, as it may seem that, again, they are not 'tough enough' for the work.

3.28

Discussion Point

Can you relate each of these five requirements to common construction site activities?

Some common construction site activities related to the requirements of the act could be the following:

- The provision and maintenance of safe plant and safe systems of work – ensuring that plant provided on site is in good working order; for example, that the forklift truck has a method statement for its safe use, a banksman may be required on certain parts of the site, and a regular maintenance schedule in place to ensure that it is working correctly.

- The use, handling, storage and transport of articles and substances – ensuring that operatives are fully trained in the materials they will be using. Handling materials should be done safely; for example, heavy items should be moved by pallet truck or other mechanical means. Storage should be well maintained so that it is not dangerous for people to gain access to their materials and that hazardous materials are locked away.

- The provision of necessary information, instruction, training and supervision – a site induction should be provided, method statements for the work which operatives are to carry out, training on new processes or equipment, for example, a PASMA ticket for using mobile aluminium towers, and a supervisor for their work.

- The maintenance of a safe place of work, with safe access and egress – safe working areas with good levels of housekeeping, no rubbish or materials in the way, safe walking routes to and from all work areas of the site, with flat level access – site management is important here.

- The provision of a safe working environment with adequate welfare facilities – safe working areas with welfare, including rest and dining areas, WCs, hot and cold running water and drying facilities.

3.29

Discussion Point

Consider the following construction activities; how can the risks be reduced through application of the hierarchy of control?

1. The movement of a forklift truck around a site.
2. Fitting edge protection to a steel beam on the edge of a first-floor slab.
3. Spraying intumescent paint on exposed soffit steelwork.

The movement of a forklift truck around the site cannot be eliminated or substituted, as mechanical means are one of the key things needed for materials management. The risk can be controlled by segregating site traffic from pedestrians with physical barriers, ensuring that the forklift has an audible siren for reversing and flashing light, and making sure that the driver is trained and qualified. Work methods could also be altered to try to offload materials from deliveries close to their final position on site to avoid using the forklift, but this may not be practical depending on the site layout.

The risks of fitting edge protection to a steel beam on the edge of a first-floor slab can be eliminated if the edge protection is fitted in the steelwork factory before the beam arrives on site. Alternatively, the edge protection can be fitted when the beam is on site and resting on the ground before it is lifted into place.

The risks of spraying intumescent paint on exposed soffit steelwork can again be eliminated by spraying the steelwork in the factory, before it comes to the site. There is then no need for anyone to work at height, or to use potentially hazardous paint in a spray form on site which may result in health issues.

Discussion Point

What TBTs might be relevant for a gang about to start excavation works?

How would you make sure that the gang is paying attention to your TBT?

Excavation works involve many different hazards. Some suitable TBTs could be as follows:

- risk assessments and method statements;
- buried services;
- working near machinery;
- working in confined spaces;
- excavations;
- supporting excavations;
- safe access to excavations;
- control of dust and fumes;
- sun safety.

You could use several management techniques to ensure that the gang is paying attention to your TBT. First, you will need to make sure you are delivering the TBT in an enthusiastic and informative way, not acting bored yourself! You will need to maintain eye contact with everyone in the gang, and make sure they are looking at you and not talking among themselves or on their mobile phones.

At the end of the TBT ask questions about what you have said to make sure that people understand. Don't just ask if they understand – the answer will be yes! – but ask how they would access an excavation or how they would then minimise dust in the area.

Discussion Point

 3.31

What do you think are the *most* important aspects of the site induction?
How would you deliver your site induction to operatives?

The most important aspects of the site induction are those *specific* to the site.

Access routes and what works are currently being carried out are important, so that workers will know *why* they may have to wear certain PPE or follow a certain walkway to their work area. This will change from site to site and even from day to day on the same site; for example, one day the crane may be working on the front of the building and the next it may be moved round to the back. The induction should change accordingly, and those on site informed through notices or TBTs.

Induction delivery must be enthusiastic and positive. If you are bored and talking in a monotone, then your audience will be bored too. You must make sure that people are aware of why they need to know this information.

If the induction is too long people will start to lose focus and not pay attention, so critical and specific information should be included rather than a repetition of standard regulations which do not differ from site to site.

Discussion Point

3.32

Wearing PPE is often a site rule.
How would you encourage people to wear the correct PPE?

Many people seem to find wearing their PPE a problem – even hard hats, which are required by law. Education is the main way of encouraging people to wear PPE. In some cases people don't realise why they have been asked to wear a dust mask, for example, and don't know the risks that dust can have in terms of long-term health damage. Telling them why can often encourage them to put a mask on.

When gloves were brought in as standard PPE there was some grumbling, but in time people realised that their hands were not getting as many cuts. There are different gloves available for many different trades, designed to suit their work, and so it might be that using the wrong PPE has put people off wearing it.

As a last resort, people must be disciplined if they are not following the rules by not wearing their assigned PPE. The rules are there to ensure compliance with legislation and so that the site can be as safe and healthy as possible. The rules must be enforced and so, in some cases, discipline is the only solution to force people to look after themselves properly.

3.33

Discussion Point

How would you describe construction industry culture?
How would you describe the safety culture on sites?

Culture is a very abstract idea; there isn't just one definition and the construction industry has had a lot of different 'labels' attached to it over the years.

A 'macho culture' is a common label, because the workforce is mostly made up of men. It has been suggested that this has also affected the safety culture – that men like to show off how strong, brave and tough they are which has led to risks being taken, and sadly accidents happening. But of course, not all men are like this, and this is a generalisation.

The industry has also been described as having a 'production-driven culture' – what matters is getting the job done, not getting it done safely. Again the need for speed and cutting costs will have had an effect on the safety culture on sites.

It has also been said that there is a 'culture of conflict'. People argue over contracts and money, over working space on site, over equipment and materials. However, this is slowly changing with new ways of working, but it may also have contributed to an 'everyone for themselves' approach which meant safety as well.

The safety culture on sites has historically been described as poor, and on some sites it still is. People work unsafely, cutting corners and taking risks just to get the job done. But on larger sites things are changing, and large organisations, with the time and money to invest, are trying to develop a positive safety culture.

3.7 Environmental management

3.7.1 Introduction

This section is all about environmental management. It aims to develop an understanding of environmental management, both the bigger picture in terms of sustainability but also how construction managers can plan and control the environmental impact of their sites.

Environmental management is a developing aspect of construction management, and there are likely to be changes to legislation and other controls for our industry in the near future to support reductions in carbon emissions and the production of sustainable developments. Good construction managers can keep up with changes in sustainability through industry journals, attending professional body CPD events and following industry news.

There are two key concepts associated with environmental management:

- Sustainability and the bigger picture.
- Environmental management on sites.

These two concepts are not actually separate. Many of the environmental management measures that can be put in place on sites will also contribute to the sustainability of our industry and the bigger picture. The term 'green' is also used to refer to the environment and is often associated with both sustainability and site environmental management practices. A green construction company would be expected to be investing in both.

We will explore the two key concepts in turn, starting with the bigger picture, so that we can then see how and why environmental management on sites can help the industry as a whole.

3.7.2 What is sustainability?

First, we need to have some idea of what 'sustainability' is. Probably the most famous definition is from the Brundtland Commission (UNECE 2005):

Sustainable development is 'development which meets the needs of current generations without compromising the ability of future generations to meet their own needs'.

This means that we need to take note of how we live today and make sure that the things we use and take for granted now – such as energy, food and shelter – are not going to limit the way future generations can live by reducing or affecting their access to the same things.

For example, we are all using more and more electricity; this book is being written using a computer, and you may be reading it on a computer, tablet or even your phone! But this energy is not without consequences; at the moment most of it comes from fossil fuels which release carbon dioxide (CO_2) into the atmosphere when they are burnt to make energy. This raises two problems. First, fossil fuels will not last for ever; they are a finite resource and so one day they will run out. Second, CO_2 is a greenhouse gas and has been linked with man-made climate change – the problems concerning this are outlined below.

Clearly the way we produce our electricity now will affect future generations in some way: there could be no fossil fuels left for them to use, and they may have to live in a world which has enhanced problems caused by man-made climate change.

Sustainability covers the practices of trying to redress this imbalance. It aims to look at new ways of producing energy through renewable resources such as wind or solar power, and it aims to reduce the amount of energy we actually use – this is often called reducing the **carbon footprint** owing to the reduction in the amount of CO_2 that will result.

This is where the construction industry has a role to play – construction is an energy-intensive industry, in the products and processes we use to produce materials and build our projects, but also in the finished projects themselves. Most of a project's energy use occurs during its operation, and significant amounts of energy can be saved by constructing more energy-efficient buildings.

Another area of industry concern in the UK surrounds the existing **housing stock**. In the UK about two-thirds of the homes we will live in by 2050 have already been built, but their efficiency is generally poor. Although new-build houses do contain significant insulation, older housing stock does not. Therefore the retrofitting of insulation and other energy-saving technologies will be needed to improve their energy efficiency and reduce their carbon footprints in the future.

A global problem

One of the major arguments for improving our sustainability comes from climate change. It has been suggested that there will be increased temperatures, and more natural disasters such as hurricanes, heatwaves, floods and drought. The Environment Agency in the UK clearly believes this to be the case, and that:

> Our houses, towns, cities, people and wildlife will need to cope with more flooding and coastal erosion, increased subsidence and greater competition for water. We will also need to find ways to keep our buildings cooler and more comfortable in hot summers, without using lots of energy to power air conditioners.

(Environment Agency 2013)

This means that we will need new ways of building, new types of buildings and more resilient infrastructure to cope with potential problems in the future.

But climate change has also made a major difference to how people, including world leaders, think about these problems. Floods do not stop at national boundaries and bush fires do not stop at passport control – the problem is a global one.

Sustainability and climate change has therefore risen to the top of the international agenda. All the countries of the world are involved, but some are keener than others to make reductions in their energy use and CO_2 production in attempts to slow global warming.

As a result there have been various international agreements and accords to try to address climate change. Several international events have been held over the past decades, including the United Nations Earth Summit in Rio De Janeiro in 1992 which produced Agenda 21, the United Nations Framework Convention on Climate Change in Kyoto in 1997 which produced the Kyoto Protocol, and in 2012 the Rio+20 United Nations Conference on Sustainable Development.

At Rio+20, the United Nations reaffirmed its commitment to Agenda 21, a document which, among other things, requires the conservation and management of resources, including water, ecological environments and the control of waste. The United Nations also urged parties to fully adopt the Kyoto Protocol, which included legally binding emissions targets for industrialised countries, including the UK.

The global focus on sustainability is ongoing, and you can find out more from the United Nations website on their Sustainable Future web pages.

Government policy and influence

'Sustainable' is one of the five aspirations for the UK construction industry as stated in the government's Construction 2025 Strategy published in July 2013.

The government has also implemented several pieces of legislation that support the international agreements and implement various European Directives. A key piece of legislation is the Climate Change Act 2008, which puts the European emissions reduction target of at least 80 per cent (from 1990 figures) by 2050 into UK law, and also sets legally binding 'carbon budgets' which limits the amount of greenhouse gases that can be emitted into the atmosphere. The Act also creates an obligation for the UK to report its progress towards this figure on an annual basis.

In order to take the first steps towards carbon reduction, the government introduced the UK Low Carbon Transition Plan which sets out how the UK will meet the carbon budgets, including the target for a reduction in CO_2 emissions of 34 per cent from 1990 figures by 2020. This plan is supported by more detailed proposals from other parts of government, including industrial and renewable energy strategies.

There are many other pieces of legislation which affect both sustainability and environmental management. These will be discussed with reference to the elements of construction environmental management that they influence as they arise.

The government also influences the construction industry in other ways. It has launched initiatives to support new ways of funding the installation of sustainable technologies, it has made changes to the planning system to encourage sustainable developments, and it has made changes to Building Regulations to ensure that sustainability is embedded in them.

3.7.3 Sustainability and construction

One of the problems in pinning down where the construction industry currently stands with regard to sustainability is that the rules may seem to be constantly changing. Many of the initiatives and changes come from government, and so if the government changes after a general election, changes in policy, regulation and guidance are often not too far behind.

As a result, this section will introduce key concepts alongside current regulations, but in order to make sure you that have the latest information and guidelines, you should always refer to original sources and government websites. Although the ideas discussed here will probably remain the same, the ways they are to be implemented may change, so again be sure to check the latest news and information.

Construction managers need to keep up to date with any policy changes and new initiatives to make sure that their projects are complying with, and making best use of, the latest developments. The best way to keep up to date with changes in this area is through industry press and professional body CPD events.

Planning policy

The government's National Planning Policy Framework (NPPF) was launched in March 2012, and states that planning approval for construction projects that are sustainable should be given without delay – what is known as **'presumption in favour of sustainable development'**.

This Framework refers to three dimensions of sustainable development: economic, social and environmental. The way the development supports these three dimensions must be considered with reference to local development plans, to assess whether proposed projects are sustainable.

The government hopes that this presumption will mean that local planning authorities can positively influence the sustainability of their local areas, and that sustainable developments can be given the go-ahead more quickly. For the construction industry this may be seen as a benefit, as it should speed up the planning process which can be lengthy, and release work more quickly.

The NPPF also sets out some clear guidance as to what development will be encouraged, and this can influence the types of projects coming online. For example, in the chapter on 'Conserving and enhancing the natural environment' the NPPF states that:

Planning policies and decisions should encourage the effective use of land by re-using land that has been previously developed (brownfield land), provided that it is not of high environmental value. Local planning authorities may continue to consider the case for setting a locally appropriate target for the use of brownfield land.

(NPPF 2012, p. 26)

Discussion Point

3.34

How could the increased use of brownfield land affect construction management?
What might construction managers need to become more aware of?

In the chapter on 'Meeting the challenge of climate change, flooding and coastal change', the NPPF requires that the:

development is appropriately flood resilient and resistant, including safe access and escape routes where required, and that any residual risk can be safely managed, including by emergency planning; and it gives priority to the use of sustainable drainage systems.

(NPPF 2012, p. 24)

Sustainable drainage systems, often known as **SUDS** or **Sustainable Urban Drainage Systems**, are those which mimic the natural movement of rainwater slowly through the ground, rather than surface water drainage systems which channel it off roofs, roads and other hardstandings, and create flood risks when discharged into rivers and streams. Construction managers will need to develop their technical knowledge and understanding of the various SUDS systems in operation. The NPPF has changed the process and criteria around planning approvals, and also encourages sustainable development practices and design. How this affects construction managers will depend on their role within the company and the project procurement route, but it has already started to influence the knowledge and skills they need to help their companies effectively plan and control construction work.

3

BIM and government soft landings

One of the driving forces behind the government's policy to encourage the implementation of BIM is sustainability. Within the BIM documentation, sustainability is defined as the need to improve the carbon performance of buildings and infrastructure, by reducing carbon outputs and improving overall sustainability.

This is also linked to the **Government Soft Landings** Initiative which looks at the whole life cycle of buildings. This takes into account the energy used during building operation; for example, the energy it will take to heat and light buildings for their entire lifetime. Being able to model building performance will influence the design and construction of projects, and ensure that sustainability is embedded in the end product, and therefore help UK developments meet the government's CO_2 reduction targets.

BIM, and its role in carbon reduction, is discussed in more detail in Part 4.

Sustainability through design

The increasing use of BIM and influences of sustainability legislation have shown that a positive way to improve sustainability is through design.

The shift in project focus – to go beyond design and construction and include operation and maintenance, and even refurbishment and demolition – has resulted in the same shift in the ideas behind design. Designs are now focused on the operation of buildings in the long term, and aim to reduce their energy consumption and carbon footprints for their entire lifetime.

An obvious example is the increased inclusion of sustainable technologies within projects. The use of solar panels on roofs and to form building façades has become more common. More recently new technological developments have included bio-active façade panels which contain living algae to provide shade and generate heat as a renewable resource.

Consideration of materials can also help a project improve its overall sustainability. Many construction products have high **embodied energy** – they use large amounts of energy to produce from raw materials, and this energy remains trapped within them. For example, the production of one tonne of cement can produce over one tonne of CO_2 (Scientific American 2008). Alternatively timber is a more sustainable material, and may be renewed over a period of time as trees are regrown. Other materials may come from overseas countries, and the energy needed to transport them to projects can be considerable. Others may be a finite resource and inherently unsustainable, or made of complex composites and therefore unrecyclable.

Selection of materials can be informed by the use of the **BRE Green Guide**, which provides an environmental impact assessment of various construction materials across their entire life cycle.

Waste can also be designed out of a project. Designing to standard industry dimensions rather than bespoke design dimensions can reduce offcuts and wasted materials, and can be helped by using BIM to enable designers to see the bigger picture of how materials will work together. Consideration of materials specifications may also minimise waste; for example, specifying a relatively soft brick will mean more damage in transport and laying, more rejects and so more waste. Using recycled and reclaimed materials can also improve the overall sustainability of a project.

Taking this even further, projects can be **designed for demolition**, where elements of the construction such as the steel frame are designed specifically to be taken apart easily and removed when the project is no longer needed, and then reused elsewhere.

Building Regulations

A key legislative tool to support the targets set in the Climate Change Act 2008 is Part L of the Building Regulations. Part L deals with the Conservation of Fuel and Power, and sets the minimum standards for construction projects.

The Regulations set requirements for the reduction of CO_2 emissions from domestic properties, which must be designed into developments, as well as requirements for airtightness. They also set U-values to be achieved by specific elements of the building. The U-value is the rate at which energy is lost through a particular type of construction. For example, the minimum U-value for a domestic roof is 0.20 $W/m^2.K$ and for a domestic wall the value is 0.30 $W/m^2.K$. The lower the U-value, the better the performance.

Part L is likely to be revised and updated on a regular basis as industry moves towards more sustainable construction, with tighter restrictions on the energy consumption of projects and increased targets for airtightness, with the ultimate goal of reaching **zero carbon**.

3.7.4 A sustainable construction industry for the future

The construction industry has a key role to play in the way society responds to the questions raised by sustainability, but this needs to be supported by a consistent approach from government.

Consideration of the whole life cycle of buildings has already started to change thinking about the role the industry plays in the longer term life of the built environment it produces. This has affected how projects are designed and planned, to ensure that they are effective and efficient in their operation and maintenance as well as their design and construction.

Our industry will need to continue to research and develop materials and technologies that can support the efficient construction of zero carbon buildings and infrastructure, as well as the methods and products to support the retrofit and modification of our existing built environment to meet the same standards.

Discussion Point

What modifications could be applied to existing domestic properties to meet sustainability criteria?

The construction industry will also need to design and construct **resilient built environments** that can survive in future man-made climate change, including extreme weather events, flooding and global warming.

It must also ensure that the materials and products used in these new structures are themselves sustainable, and may be reused or recycled without detriment to future generations.

For construction managers, what originally started as site management, and extended through design and build, now includes consideration of the maintenance and operation of projects for their entire life spans. This has created a long-term sustainability perspective and construction managers now need to consider generations to come, and how their actions on site have much wider repercussions on the global stage.

There is far more information available around sustainability and construction than can be included here. It is a growing way of thinking, and is starting to affect everything we do as an industry, as well as the way we as a society will live, work and play in the future. Keep up to date with industry CPD events, or begin to explore further through the websites listed at the end of this section.

3.7.5 What is environmental management?

If sustainability thinks about the bigger picture and looks at the environment on a global scale, environmental management aims to manage, control and mitigate the impacts of industry activities and practices at the local scale of the construction project and site.

This includes concern with the effects of projects on the local ecology, the quality of life for the local residents during construction works, and the prevention of any pollution within the local environment. It also means that construction managers should monitor materials used and aim to reduce waste where possible on their projects, as well as to make sure that resources are being used effectively, such as water and energy. But this also influences the bigger picture and forms part of sustainable construction.

For example, if the construction manager sets out to minimise energy use on their project by improving the way plant is managed and maintained on site, not only will their site reduce its carbon footprint, but they will also help reduce the carbon footprint of the UK industry. Environmental management focuses on the small things that, when all of the UK construction sites are added together, can make a major difference to the sustainability of the UK construction industry as a whole.

The main regulatory government body for environmental management in the UK is the **Environment Agency** (EA), which is tasked with the protection and improvement of the environment and the promotion of sustainable development. It also ensures compliance with and enforcement of environmental legislation.

The EA is consulted by local planning authorities on planning applications, and can therefore influence the planning conditions placed on a development.

Among other issues, the EA is concerned with the following:

- protecting and improving air and water quality;
- flood mitigation;
- protection of wildlife;
- management and remediation of contaminated land;
- waste management and control;
- reducing climate change.

All of these issues are also relevant to the construction industry and construction management.

3.36

Discussion Point

How could construction negatively influence each of the above? What common construction practices could have an effect?

The EA is able to provide support and guidance for the best way to mitigate any impacts construction activity can have on the environment. The EA can be contacted before development starts to provide advice on the proposals and design of the project, as well as information to support effective and efficient environmental management on site.

As the EA (2014) says:

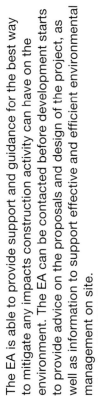

Good site layout, resource management, and taking sensible steps to minimise pollution can greatly reduce the environmental impact of your site.

You can find out more about good site layout, with full reference to environmental management, in Section 3.2.3.

The contact details for the EA, as well as more information about what it does, may be found on its website.

3.7.6 Environmental management tools

There are a variety of tools available to help support environmental management on site:

- **BREEAM** provides a way of measuring the 'green credentials' of a project through a BREEAM rating, and sets out goals for construction managers to achieve on site.

- **Environmental Impact Assessments (EIAs)** are a legislative requirement for some projects, but they are also used as a check for other projects.

- ISO 14001 provides international standards for environmental management through the development of **Environmental Management Systems (EMSs)**.

- **Environmental Management Plans (EMPs)** are project-specific documents, produced as part of an EMS.

The extent to which construction managers are involved in these elements will depend very much on the type of project, its scope and its procurement route. While they may be involved in the planning of the EMS or help to undertake an EIA, they will more likely be made responsible for the planning, development and implementation of the site-specific EMP, and meeting the BREEAM management requirements on site. Whatever the role of the construction manager, they need to understand the different tools available and any necessary requirements in order to make sure environmental management is efficient and effective on their project.

BREEAM

In order to prove a measure of sustainability has been achieved by a project, a rating system known as BREEAM, may be applied which grades the project as either Pass, Good, Very Good, Excellent or Outstanding, and awards it a star rating from 1 to 5.

BREEAM stands for the **Building Research Establishment Environmental Assessment Method** for construction projects. It is a certification system for the environmental performance of buildings, established by a comprehensive review of the design, construction and operation of projects carried out by an approved BREEAM Assessor.

Many construction clients want a certain BREEAM rating for their project in order to ensure a sustainable future for their new building, and will make sure that their design team work to achieve this goal. Many developers also aim to construct their projects to a high BREEAM rating, and this can benefit their sales and lettings, as end users can be reassured that their accommodation will be environmentally friendly and energy bills should be lower than normal.

There are several different assessment schemes under BREEAM, but they all evaluate their projects using a range of criteria against which credits can be awarded, and are added together to ultimately give a final score for the project. For the New Construction 2011 Scheme these criteria are as follows:

- Management
- Health and well-being
- Energy use
- Transport
- Water use
- Pollution
- Materials
- Waste
- Land use and ecology
- Pollution
- Innovation.

(BREEAM 2012)

The first of these criteria is directly linked to construction management. The management category (MAN) contains two subcategories (MAN02 and MAN03) which are closely linked to environmental management on site.

First, category MAN02 asks for evidence of responsible construction practices, which BREEAM defines as those that are managed in an environmentally and socially considerate, responsible and accountable manner. This may be demonstrated by compliance with the Considerate Constructors Scheme, more details of which may be found in Section 3.2.4.

In category MAN03, BREEAM awards credits for mitigation of construction site impacts and to recognise sites managed in an environmentally sound manner in terms of resource use, energy consumption and pollution. To demonstrate this, water and energy consumption should be monitored and recorded, the use of transport for construction materials and waste to and from the site should be noted and mitigated, all timber used on the project must be from sustainably managed sources, and there should be good practice procedures for pollution prevention.

Discussion Point

How could you practically monitor the transport aspect of MAN03? How could you confirm that all timber was sustainable?

Construction managers are usually tasked with the achievement of these two categories within the BREEAM assessment of a project, and will need to implement the necessary processes and procedures to meet them.

You can find out more about BREEAM, including details of the schemes and the specific criteria around environmental management including MAN02 and MAN03, on its website.

Environmental impact assessments

Environmental Impact Assessments (EIAs) are a legislative requirement under the **Town and Country Planning (Environmental Impact Assessment) Regulations 2011** for certain types of developments and relate to the environmental impact of completed projects on their surroundings, rather than any construction works.

The EIA process results in the production of an **environmental statement** (ES), which summarises the development, explains what other proposals have been considered, why this one has been selected and what mitigation measures will be put in place to reduce the impact of the project on the environment. The EIA and ES must be submitted with the planning application, and will be used by the local authority to assess the planning application and require approval before permission will be granted.

Some clients and contractors voluntarily undertake EIAs even when not required to do so by law, to see how their proposed project will affect the local area, and how they could develop the design or change their proposal to reduce its environmental impact. This is a way through which both clients and contractors can demonstrate a level of commitment to corporate social responsibility (CSR).

For example, the international contractor Lend Lease makes mention of an EIA process in its **Environmental Policy** Statement:

> We will protect biodiversity and land quality through the ongoing assessment and management of our impacts.
>
> (Lend Lease 2009)

Environmental management standards

ISO 14001 is the internationally recognised standard for the management of organisational environmental impact for all industries, not just construction, as developed by the International Organisation for Standardisation.

It standardises the production of environmental management systems (EMSs), which enable construction companies to identify their environmental impacts, put effective controls in place to manage them, and ensure that all practices and processes meet current legislative requirements.

Many construction companies are ISO 14001 accredited to demonstrate their commitment to environmental management. For some projects ISO 14001 accreditation can be a pre-tender

Using a type of EIA on all projects, not just those that require it by law, means Lend Lease can show commitment to the environment, as well as measure this commitment through the steps it has taken to minimise environmental impact through the EIA process.

requirement, as it reassures clients that their project teams are serious about sustainable construction.

Environmental management systems

Environmental Management Systems (EMSs) aim to create systems through which companies can ensure that they are meeting its environmental commitments and legal obligations.

This process starts with the creation of an environmental policy to set out company goals and objectives. The following statement is from BAM Construct UK's Sustainability Policy:

> Environment: We will limit our environmental impact. We take all reasonable measures to ensure that our activities are conducted in a way that minimises our impact on the local environment. We promote good environmental practice and seek opportunities to enhance biodiversity on our construction sites.
>
> (BAM Construct UK 2011)

The EMS is then developed to provide the framework through which the policy may be implemented, and this should follow the ISO 14001 guidance.

This includes the identification of those responsible for the implementation of the EMS, who will need what type of training to support the EMS, and how the EMS will be communicated. It may also include the development of a set of environmental documents to standardise the environmental management process, such as forms to be used for weekly environmental inspections or a template for site Environmental Management Plans.

Smaller companies may not have the resources to put a formal EMS in place or attain ISO 14001 accreditation, but there are benefits to adopting an organised and consistent company approach towards environmental management, not least to make sure that all legislative requirements are met. For many companies, good environmental management can be achieved through detailed and well-structured Environmental Management Plans.

Environmental Management Plans

The Environmental Management Plan (EMP) is the project-specific document that clearly sets out how environmental management is to be carried out on the construction site. It includes any necessary elements of the company EMS, if they have one, the legislative requirements for the project, as well as addressing any environmental issues that may have been made **planning conditions** by local authorities when granting planning permission for the site.

An example of an actual planning condition is as follows:

Prior to the commencement of the development the developer shall submit a Dust Management Plan for the written approval of the Local Planning Authority. The Dust Management Plan shall identify all areas of the site and site operations where dust may be generated and further identify control methods to ensure that dust does not travel beyond the site boundary. Once in place, all identified measures shall be implemented and maintained at all times. Should any equipment used to control dust fail, the site shall cease all material handing operations immediately until the dust control equipment has been repaired or replaced.

In order to discharge and sign off the condition, the local planning authority requires a Dust Management Plan to clearly show how dust will be managed on the site and how it will be controlled throughout the project.

Such planning conditions are common, and often also refer to noise, vibration or light pollution. The condition requirements may depend on the surrounding area. The above condition was from a project being carried out in a residential area where construction site dust could quickly become a problem for site neighbours.

All of the planning conditions and other site environmental management processes are brought together in one project-specific EMP. An EMP will likely contain the following information:

- Outline of the project.
- Site team and responsibilities.
- Hours of work.
- Traffic management.
- Environmental impacts and mitigation proposals:
  - noise;
  - dust;
  - vibration;
  - light;
  - air quality;
  - prevention of pollution;
  - contaminated land.
- Waste management.
- Ecological protection measures (including tree protection orders).
- Incident management and response.
- Communication of the EMP.
- Plans for review and revision.

Other sections may be needed depending on the type of project, the location, environment and other planning conditions placed on the project by the local authority. Although EMPs can follow a template, often contained in the EMS, they must be site specific to ensure that the environmental impacts of that particular project are addressed.

3.7.7 Environmental management on site

This section examines in more detail the practical ways in which environmental management is carried out on sites by construction managers. It aims to provide practical details so that construction managers can not only comply with the rules, but also implement good practice through effective planning and control.

Environmental management can be suggested to cover the management of the following three areas:

- What is already on the site.
- What is consumed by the site.
- What is emitted by the site.

These three areas help construction managers focus on how the environment may be affected by our sites and work processes, and how best to mitigate or even prevent any impacts.

What is already on the site?

An example site is shown in Figure 3.7.1.

This site once contained buildings that were demolished a number of years ago, and now vegetation has grown over the rubble left on the site. A housing development is now proposed with the construction of 120 dwellings.

This is a brownfield site – it has previously been used for development. This means that there is the possibility that the site contains **contaminated land**, either from previous use (if the site was

Figure 3.7.1 A site ready for development

used for industrial purposes there may be contaminants in the ground; for example, machine oils or dyes), or from the building that has been demolished – depending on its age, asbestos may have been present in the structure. This can be determined by the **Desktop Study** for the project, which in this case shows that there were university buildings previously on the site.

Discussion Point

3.38

How could you find out the previous land use of a site?
Where would you visit for information?

As the demolition was carried out recently, all asbestos should have been removed from the buildings before their demolition; however, there is the chance that other contaminants may still be in the ground from workshops or other light industry on the campus.

Professional consultants should be involved from the **feasibility** stage of any project, and can provide guidance and undertake the necessary tests to confirm whether the ground is contaminated. A full risk assessment needs to be carried out, and different options to solve the problem examined before a final remediation strategy is proposed. More details may be found in the *Guiding Principles for Land Contamination* produced by the Environment Agency and available through its website.

Contaminated land can pose several risks:

- Risks to the local environment: if pollutants within the ground are allowed to travel they may cause damage elsewhere. For example, rain falling on disturbed ground could pick up pollutants as it leeches through to the **groundwater**, and poison plants and trees in the local area as it travels past them to the local river.

- Risks to the health of the construction operatives working on the site: if they are dealing with the ground as part of their work, they could be affected by contact or airborne contaminants.

- Risks to the health of the new residents: these could be in the form of residual pollutants in the contaminated land that may leech to the surface over time, and again cause health issues for those now living on the development.

Therefore it is essential that contaminated land is managed correctly to avoid any repercussions for the workforce, the new residents and the local environment.

Another consideration about the site is its existing ecology. Our site has some mature trees, as is shown in Figure 3.7.2.

Trees can be protected by **Tree Preservation Orders** (TPOs), which are allocated by the local planning authority due to the age, size or significance of the tree to the local area. A TPO makes it an offence to remove or damage a particular tree without permission. Details of all trees in an area covered by TPOs may be obtained from the local planning authority, and the large trees shown in Figure 3.7.2 have TPOs on them.

Figure 3.7.2 Mature trees on the site

Any trees with TPOs, or other existing trees that are to be retained as part of the development, must be protected during the construction works. They should be securely fenced off to ensure that they are not accidentally struck by construction vehicles, but more importantly to protect their roots from damage caused by loading the ground. For example, if heavy materials are stored under trees, this could damage the root systems and even kill them.

A planning condition was made on our site about the trees in place. It states:

No development shall be started until the trees within or overhanging the site which are the subject of a Tree Preservation Order have been surrounded by fences of a type to be agreed in writing with the Local Planning Authority. The approved fencing shall extend to the extreme circumference of the spread of the branches of the trees (in accordance with BS 5839) or as may otherwise be agreed in writing with the Local Planning Authority; such fences shall remain until all development is completed.

In order to discharge this condition, a protected zone should be established using the specified fencing. Signs should also be used to explain why the area has been fenced off. A further condition requires the inspection and approval of this fencing once it is in place.

Trees, **hedgerows** (which are also protected by law) and other vegetation on the site may also contain wild birds and animals.

Identification of any **native species** of birds and animals on the site must be established by a full **ecological survey** carried out by consultants at the feasibility stage of the project. To make sure the survey captures all the native and protected species, a number of visits will be needed at different times of the year and during both day and night to establish what species are present and where they are located. This needs careful and early planning to avoid delays later in the project. If ecological surveys are not carried out they can be made a planning condition, and it may be months before the correct time to survey comes round again.

Any birds found to be nesting in trees when work begins on site cannot be disturbed by law, under the **Wildlife and Countryside Act 1981**. Therefore any trees or vegetation that need to be removed for the new housing development must be taken off site before the birds start to nest. This should be planned for during the winter before the birds start to nest, or late summer and autumn when the baby birds have flown. In England the nesting season is between 1 March and 31 July, as advised by **Natural England**, the agency responsible for the management of the local environment.

This rule also applies if birds have set up home in derelict buildings that are to be demolished, or even new buildings under construction. The birds and their nests are protected, and so construction teams must wait for the young to leave the nest before they can start work. It is important to secure projects wherever possible from wild birds to prevent them from nesting in the structures.

Surveys may also identify **protected species** on the site, which are those as defined by the **Conservation of Habitats and Species Regulations 2010**, as well as birds listed within Schedule 1 of the Wildlife and Countryside Act 1981. Protected species include badgers, bats and great crested newts.

In order to work on projects which will impact upon any of the above, licences are needed to capture or disturb them, or to destroy any of their habitats. Any activity should involve the help of specialist consultants who are able to advise and carry out the work correctly. Licences are issued by the Wildlife Licencing Unit, part of Natural England.

Surveys should also confirm whether any **invasive species** can be found on the site. These are plants not native to the UK and whose speed of growth causes problems for our own native species which struggle to survive alongside them.

Common species that affect construction sites are as follows:

* Himalayan balsam
* Giant hogweed
* Japanese Knotweed.

Japanese knotweed has been found on our site along one of the boundaries, a stem can be seen in the centre of Figure 3.7.3. Japanese knotweed can grow a metre in height in one month and this growth is also very strong; the plants can damage building foundations and even grow through paving and tarmac. It is an offence under the

Figure 3.7.3 Knotweed on the site

Wildlife and Countryside Act 1981 to cause the spread of Japanese knotweed, and so construction companies are legally required to make sure that they manage or remove the plants effectively.

Knotweed does not produce seeds but grows from rhizomes (parts of its stems or roots) and so can be easily spread by small pieces of the plant sticking to people's boots and machinery wheels or tracks.

Discussion Point

How would you prevent the spread of knotweed on the site? What practical measures could you take?

The removal of Japanese knotweed is a specialist task and potentially very expensive, as any waste produced from the plants must be managed carefully and in accordance with the Environment Agency Code of Practice. It may be treated with chemicals at certain times of the year by certified professionals, although it can take three years for the plant to stop growing back. It may also be physically cut back and removed through a bulk dig process. There are also options to bury or burn the plants on site, but this will require permission from the Environment Agency. For more information, please refer to *The Knotweed Code of Practice* on the EA website.

Ecological surveys should establish what is already on the site. The findings may then be used to plan the design and management of the project to minimise any impact upon the local environment. The construction manager may become involved with the management of issues during the construction phase of the project, such as the protection of any identified habitats, any TPO trees on the site, and how the knotweed removal will be carried out under strict controls. All this information should be contained within the project's Environmental Management Plan.

What is consumed by the site?

During the construction phase of projects, sites will use up different resources as the work progresses, and from a sustainability perspective this includes the consumption of energy and water. If the construction manager can reduce the use of these resources on the site, it will help improve the sustainability of the construction industry as a whole.

Reducing the energy consumption of sites reduces their carbon footprint – the total amount of CO_2 emissions produced by sites during the construction of projects.

Many large contractors are now focusing on energy consumption and cutting carbon. Laing O'Rourke (2013) says:

We have set ourselves the target of cutting our carbon footprint by 30 per cent by 2020 (against our 2008/09 baseline). By 2050 we aim to achieve an 80% reduction. Working with independent environmental bodies and government regulators, we are actively taking steps towards these targets.

The Environment Agency has produced a construction-specific **carbon calculator** which can assess the carbon footprint of a site, and highlight areas of significant energy consumption on the project. The calculator includes the materials to be used in the construction, the amounts to be used, the transportation costs and their embodied energy. For example, the cement in concrete has a high level of embodied energy, but this may be reduced by using recycled aggregate in the mix or even other materials such as recycled car tyres.

The calculator shows which materials are increasing the energy consumption of the project and allows the construction manager to target any reductions. For domestic projects this could mean a change in the concrete specification for the house foundations and slabs once any alternative materials have been checked to ensure that the concrete will still meet the quality requirements of the build.

On site, the construction manager should focus on the efficient operation of plant and equipment to ensure that energy in the form of fuels and oils is not being wasted. For fixed plant running on electricity, such as tower cranes, meters may be used and diesel use monitored for mobile plant. Regular maintenance of all plant on sites will ensure their efficient running for the project duration.

Water is another natural resource, and needs careful management and monitoring to ensure that it is not being wasted on sites. Water is often metered by volume of use, and so there can be cost benefits to saving water as well.

Large contractors are also keeping an eye on their water consumption. In its Environmental Policy Statement, Lend Lease (2009) says:

It is our long-term aspiration that all our business operations and all buildings we produce and/or operate are zero net water as a minimum. We will therefore continue to reduce water consumption, by improving the efficiency of water use in our activities and developments, including through recycling.

The construction manager should carefully plan the provision of water to where it is needed through underground temporary water connections and stand-pipes, to remove the need for long hose connections that often leak. Making leak spotting part of the weekly inspections and daily briefings can ensure that if a leak or waste of water occurs, it is quickly stopped. Thinking about possibilities for recycling can also reduce overall consumption.

The **Strategic Forum for Construction** (2012) has produced a guide to saving water on sites. The guide spells out key ways in which construction managers can reduce water use, including how to use water efficiently where necessary; for example, making sure that **dust suppression** spraying is effective, or using **wheel washes** which recycle the water as they wash the muddy vehicle wheels when they leave the site.

BREEAM Assessments examine both water and energy use of projects through monitoring and recording consumption, as noted in Section 3.7.6.

Discussion Point

The site office and welfare cabins can also use a lot of energy and water.
How could you reduce consumption in these areas?

The construction manager must plan the use of energy and water on their site, and ensure that both are managed and monitored. Making energy and water consumption an agenda item at the weekly site meetings, and providing training and regular tool-box talks, can make sure that the whole site team keeps focused on these important environmental issues and avoids over-consumption of resources on the project.

What is emitted from the site?

As well as the production of the finished project, construction sites can produce many other things, including environmental nuisances, such as dust or noise, pollution and waste.

Environmental nuisances are as detailed in the **Environmental Protection Act 1990**, and must be mitigated where possible; these include the following:

- fumes;
- dust;
- noise;
- vibration;
- light.

All of these nuisances are produced in some way or another by construction activity, but can be reduced and even removed with good site management. Often their management during the construction works is made a planning condition, as noted in Section 3.7.6, and should form part of Environmental Management Plans.

Such reduction or mitigation methods may include those shown in Table 3.7.2.

It is the role of construction managers to ensure that any mitigation controls put in place are used correctly, are monitored to ensure their effectiveness, and are revised as and when required.

Table 3.7.2 Environmental Impacts and Mitigation Measures

Environmental Impact	Possible Mitigation Measures
Fumes	Good maintenance of plant and machinery. Positioning of fixed plant (such as diesel generators) away from sensitive building.
Dust	Sheeting of loaded vehicles leaving site. Site speed limit restrictions to avoid raising dust. Use of jet and wheel washes where necessary.
Noise	Restricted hours of working for noisy activities. Location of noisy plant or equipment away from sensitive buildings.
Vibration	Continuous monitoring of vibration around the site. Change in work methods where possible (e.g. use of continuous flight auger piling rather than driven).
Light	Shut off floodlights at night. Ensure they are directed at the work area and not causing light pollution to neighbours.

More serious emissions from sites can result in pollution; the EA estimates that several hundred pollution incidents occur on construction sites each year. In response, it has produced the **Pollution Prevention Guidelines (PPG6) for working at construction and demolition sites.** This is a practical and detailed document, and provides a great tool to help construction managers in fulfilling environmental management responsibilities.

Prevention of pollution requires good environmental management planning, and potential polluting activities should be identified through EIAs or within EMPs. The **Polluter Pays Principle** of environmental law means that whoever causes the pollution will have to pay to make good any damage to the environment that such incidents cause, so it is essential that the planning processes capture all potential issues. In addition, serious incidents can result in custodial sentences and large fines for those involved.

One of the most common forms of pollution from sites comes from **water run-off.** Construction work creates a great deal of dust and mud, especially during groundworks and excavation activities. These activities all produce silty water run-off, which in turn can block drains and cause problems in local streams and rivers, killing fish and plants. Other chemicals from spills, oil or diesel leaking from plant, or rainwater leeching through waste skips may also be included in the run-off, making it a potentially very hazardous pollutant.

Discussion Point

3.41

What common construction activities could produce run-off? How could you plan to contain it?

On brownfield or redevelopment sites, locating the surface and foul water systems should be the first step to making sure that no polluted water is accidentally discharged into either. On greenfield sites the local rivers and streams should be identified for the same reasons.

Permits will be required from the EA for any discharge from site into surface water systems including rivers and streams, and from the local water company if discharge is into foul sewers. In some cases treatment may be required, or settlement tanks or Sustainable Urban Drainage Systems (SUDS) such as **swales**, to remove any silt before it is discharged.

Pollution may also be caused by poor maintenance of plant and equipment, which can leak fuels and oils. The construction manager must ensure that maintenance and inspection systems are in place for all plant on the site to control these potential pollutants.

Other issues may arise during refuelling, and **drip trays** should always be placed under fuel tanks to capture any drips. Drip trays are large plastic or metal trays which capture leaking fuel, preventing the pollutant from reaching the ground.

SECTION 6: ACCIDENTAL RELEASE MEASURES

6.1. Personal precautions, protective equipment and emergency procedures

Risk of contact: Wear protective gloves and goggles/face shield.

6.2. Environmental precautions

Do not discharge into drains, water courses or onto the ground. Contain spillages with sand, earth or any suitable adsorbent material.

6.3. Methods and material for containment and cleaning up

Wear necessary protective equipment. Extinguish all ignition sources. Avoid sparks, flames, heat and smoking. Ventilate. Absorb onto sand, earth or absorbent granules. Pick up with non-sparking tools and place into metal containers with sealable lids. Mark containers with details of hazards and store in designated area ready for disposal.

Figure 3.7.4 Extract from Bostik's safety data sheet for bitumen paint

Emissions of any hazardous materials can result in pollution. Hazardous materials are common on construction sites, including diesel fuels, bitumen paints and epoxy resins. Construction managers should be aware of what materials are present on their site, and the method statements and risk assessments for their use should include the **safety data sheet** for the material, to explain how it should be managed, and how to control any accidental spill or release. If possible, construction managers should try to substitute for non-hazardous materials, thereby removing the risk from their sites.

For example, Figure 3.7.4 shows an extract from Bostik's safety data sheet for bitumen paint (2011), about what to do in case of accidental release.

3.42

Discussion Point

What would you need to plan for if this material was to be used on your site?

How would you communicate this to the site team?

Spills should be tackled by trained operatives using the appropriate **spill kit** for the type of spill. This could involve using special granules to soak up the liquid, or cotton socks and pads to contain and absorb it. All waste including the spill kit materials becomes hazardous after use and will need to be disposed of accordingly. A spill kit is shown in Figure 3.7.5.

The best way to manage accidental spills is to avoid them by ensuring good materials management and maintenance practices. Training and education can be some of the most effective measures in preventing spills on sites, and holding regular TBTs will make sure that everyone is aware of the polluting materials in use on the site, and what to do and who to contact in case of an incident.

Another product of construction sites that has an environmental impact is waste. Waste affects both environmental management and sustainability. Wasting natural resources and the energy that has gone

Figure 3.7.5 Spill kit on site

into producing materials negatively affects the sustainability of the project, while physical waste can also cause environmental problems as a source of pollution and landfill.

Construction managers should approach waste management through the **waste hierarchy** set out in the **Waste Regulations 2011**. These Regulations state that waste should be dealt with in the priority order of the following:

- Prevention (avoid its production).
- Preparing for reuse (clean, refurbish and reuse as is).
- Recycling (turning the waste into a new product).
- Other recovery (for example, burning with energy recovery).
- Disposal (landfill or incineration with no energy recovery).

Following the waste hierarchy, waste management should start with attempts to prevent its production, before work even starts on site.

Discussion Point

How can construction managers prevent the production of waste on sites?

What areas of site management should be prioritised?

The next step in the hierarchy is to try to reuse the waste. Some larger companies have set up their own licensed waste management operations to take waste from their sites for assessment, grading and reuse. This saves in costs, such as **landfill taxes**, and also produces more sustainable materials for use on the next project. For example, bricks can be salvaged, tested to ensure they meet BS standards, and reused in a new construction project. Alternatively, they can be crushed and mixed with other recycled aggregates to form products that may be used in pavement or road buildup, depending on the strength of the final mix. In some cases, exemptions and **permits** apply to the use, treatment and storage of waste within the construction industry and must be obtained prior to the activity taking place.

Working down the hierarchy, waste can also be recycled into other materials and products. On sites, waste stations should be established with different skips used for different materials, segregating timber, metal and plastic for recycling. Some registered waste carriers will collect mixed skips (not including hazardous waste) and segregate them at their depots for reuse and/or recycling. Construction managers have a duty of care to ensure that their waste carriers are licensed (the EMP should contain the names and licence numbers of all waste carriers used by the site) and a waste transfer note is correctly completed for each consignment of waste removed from site for recycling.

Some construction waste is classified as hazardous, and its management must comply with the **Hazardous Waste Regulations 2005**. Hazardous waste should be kept in closed and sealed containers to prevent pollution. Sites which will produce over 500kg of hazardous waste must register with the Environment Agency. Many construction materials are classified as hazardous, such as paints, silicones and coal tar from road arisings, so sites often have to register as a matter of course. Construction also produces other specialist wastes; for example, plasterboard and other gypsum products cannot be sent to landfill mixed with other waste.

For waste that cannot be recycled, the final options are either to seek other recovery, which can include incineration by specialists to produce energy, or to simply dispose of it as mixed waste to landfill on which landfill tax must be paid.

Waste management is specifically legislated for in a number of regulations, and failure to comply with the law risks penalties in court and may lead to serious impacts on the environment. A large amount of fly-tipping contains construction waste, and this should be avoided by ensuring professional management of waste at all times.

With increasing focus on sustainability, it is not surprising that regulations around waste and waste management are changing and have already become more restrictive in attempts to encourage the reuse and recycling of resources. They are likely to keep changing in the future and it is essential that construction managers keep up to date with the latest developments through CPD to ensure that their sites are compliant.

You can find out more about waste management on the government's **WRAP** website, which also has a specific construction industry area.

Checking and control

Like all aspects of construction management, environmental management is not a one-stop activity. Plans are not put in place at the beginning of the project and then left to their own devices. Environmental management is an ongoing process and requires management control to ensure that the plans and mitigation activities are being carried out correctly, and that the existing provisions remain the most effective and efficient for the project at any given stage.

Environmental management should form part of the standard weekly site inspection, with special attention paid to the preventive measures put in place. For example, plant maintenance records should be checked to ensure that they are up to date and spill kits regularly inspected to make sure they contain the necessary equipment to prevent a potential pollution incident caused by work in the immediate area.

Discussion Point

What other environmental management measures could be considered?
How would you check and control them?

More detailed environmental inspections should also be undertaken on a longer term basis; for example, the Environmental Management Plan should be reviewed monthly to check that all required measures are still in place, and if any new measures need to be brought in due to new works.

Where problems have been spotted, for example, if waste is not being controlled correctly or there have been violations of areas segregated for ecological reasons, construction managers must take action to resolve the issues. Putting things right is the first course of action, and then educating the workforce to make sure that they are aware of the reasons behind the control measures that have been put in place. As a last resort, disciplinary procedures may be required.

As within any form of construction management control, ongoing monitoring of the environmental management should form part of the construction manager's daily activities. General site walk-rounds will quickly show an observant construction manager whether the site is keeping up to date with the practices needed to make their site as sustainable as possible.

3.7.8 Corporate social responsibility

Many construction companies consider themselves to be sustainable and environmentally friendly, and these ideas and the actions associated with them are often included within the wider concept of corporate social responsibility (CSR), which is also known as 'Social responsibility', just 'Responsibility', or even 'Sustainability'.

The idea behind CSR is that as well as working for their own gains, such as profit or expansion, companies also have a responsibility to the community in which they do their work.

Seddon Construction, a long-established contractor in the North West of England, have stated on their website that:

> We are responsible for ensuring we minimise that impact by being accountable for our actions and by responding to the different concerns and demands of our customers and communities we engage with, while remaining profitable and competitive. This means conducting our activities according to rigorous ethical, professional and legal standards. In this way, Seddon continues to deliver high-quality services that help ensure a sustainable environment for future generations.
>
> (Seddon 2013)

This statement includes ideas of responsibility and accountability, as well as sustainability.

Companies are starting to behave in a more responsible manner which can also involve giving something back to the community in which they are working. This means that social and community programmes are often organised around large construction projects; for example, an educational school trip around the site or employment of apprentices from the local area both count as contributing to a company's CSR.

The aim of CSR is to support the community, and the company, over the long term, and research has shown an association between investment in CSR and good financial performance, although no causal link has been proved.

Discussion Point

Why do you think CSR could help profits?

CSR is also being driven by industry clients, who are able to include CSR requirements within the tender documents for a project. Local councils often take advantage of this, and can set targets for local labour employment, school liaison or the support of other neighbourhood events such as providing spare topsoil for allotments or helping to renovate old buildings using materials from sites. Contractors who can demonstrate a commitment to CSR strengthen their prospects of winning work.

Society has also come to expect a level of CSR in the way companies operate, and construction companies are no different.

There is also growing pressure from governments, international bodies, investors, charities and other regulators for improved accountability, which is in some cases being supported by legislation.

But CSR can also attract criticism. CSR projects must be carefully considered and go right through the supply chain.

Discussion Point

3.46

What problems could arise if CSR is not embedded in the supply chain?
How would you make sure it was?

Sometimes CSR is considered to be 'greenwash' with companies saying they are doing things when in reality they are not. CSR should be easily measurable and reportable, and as it involves interactions with many different stakeholders it can't always be positive news!

CSR and sustainability are about the same thing from different perspectives. CSR seeks to satisfy those the organisation is *responsible to*, including stakeholders, such as shareholders, employees and the community. Sustainability concerns issues which the organisation is *responsible for*. But these two aspects are often meshed together, as the concerns of stakeholders are often about issues for which the organisation is responsible.

CSR is now common practice within the construction industry, and all large companies have a tab on their website linking to CSR, Sustainability or Responsibility. Many companies are proud of their CSR and promote their CSR activity alongside their biggest and best completed construction projects. It is also helping to tackle the poor image of the industry, and poor relationships with its neighbours. CSR requires stakeholder engagement, and this includes the local community. Engagement in the form of early involvement in projects for all stakeholders should mean fewer issues during the construction phase. For the construction industry this can be one of the most important aspects of CSR and one way to reduce problems for the construction manager on site.

CSR is changing the way the industry thinks about environmental management and sustainability. It is no longer a 'bolt-on' to the practice of making money, rather it is seen as a way to support work winning and develop a good reputation. Companies are keen to demonstrate their environmental and sustainable credentials, and to become involved in their local communities, which can only benefit neighbourhoods and local wildlife as well as generations to come.

3.7.9 Environmental management: summary

This section has looked at environmental management from two perspectives.

First, the global perspective, the bigger picture of global sustainability and the role the construction industry has to play in ensuring that future generations are not compromised by the way we are building today. Sustainability can be embedded in construction activity throughout project development, through the project design and use of new and sustainable technologies.

Sustainability has also been supported by the UK government through changes to planning policy and the implementation of BIM, but although some government initiatives have helped to ensure that the industry moves towards a more sustainable future, some have not helped the sustainability of the industry itself, and changing legislation and stalled initiatives have caused problems.

Second, the local perspective focused on the construction site and the potential impacts upon its immediate environment, as well as the way in which these local impacts can also influence and contribute to the bigger picture of UK and global sustainability.

Environmental management tools are available to support construction environmental management, and guidance on all areas may be sought from the Environment Agency and Natural England. EIAs, EMSs and EMPs may be used by construction managers to assess the environmental impacts of their projects, and plan for mitigation.

Initially this involves the management of the existing environmental aspects of the site before work commences, such as protection of the local ecology and management of contaminated land. Planning and control of the operations of the site will ensure that resources of water and energy are not wasted during construction processes to reduce the site's carbon footprint. Control of emissions from the site is also vital during the construction phase, to reduce nuisances, pollution and waste.

The construction industry is developing and enhancing new approaches to environmental management and sustainability, including the implementation of CSR. It is now an integral part of how companies operate; over and above complying with legislation they now go the extra mile to protect, enhance and help improve the local communities in which they work.

Discussion point comments

Discussion Point 3.34

How could the increased use of brownfield land affect construction management?
What might construction managers need to become more aware of?

The increased use of brownfield land will mean that construction managers need to be more aware of environmental management practices, such as how to effectively manage contaminated land and keep their workforce safe and healthy in such environments. As more brownfield sites are developed, those that are left are likely to be the more challenging in terms of the pollution control and management needed.

Discussion Point 3.35

What modifications could be applied to existing domestic properties to meet sustainability criteria?

Modifications to existing buildings are also known as retrofit works. On domestic properties energy use may be reduced by the use of alternative methods of energy production such as **photovoltaic panels** to generate electricity, or solar panels for hot water – both may be fitted to south-facing roofs.

Improvements in insulation, which reduce energy use, can be made in a number of ways, either through insulation fill within the cavity walls, or through insulation panel systems fixed to the exteriors of properties. Loft insulation can be fitted or increased, insulation fitted to heating pipe work (lagging) to reduce heat loss, and double-glazing installed.

Discussion Point 3.36

How could construction negatively influence each of the above?
What common construction practices could have an effect?

Please see Table 3.7.1.

Transport is often monitored on large sites by the use of a log kept in the gatehouse to the site. All delivery drivers, skip wagon drivers and visitors to the site are required to log in with security and, alongside their name and company name, the location where they started their journey is noted and where they will be returning to. This information

Table 3.7.1 Construction influence on environmental commitments

EA commitment	Construction influence
Protecting and improving air and water quality	Dust and fumes can have a negative impact on air quality in the local area, this can be caused by plant and site traffic. Water quality can be negatively affected by run-off containing silt, this can be caused by damping down or concrete wash-outs.
Flood mitigation	Silty water can block drains and cause localised flooding. Significant levels of run-off can also have a negative effect downstream.
Protection of wildlife	Removal of habitats can affect local wildlife, trees and shrubs may need to be removed for the planned development to go ahead. If no surveys are carried out this can impact local wildlife.
Management and remediation of contaminated land	Developing contaminated land without correct remediation can cause pollutants to affect the local.
Waste management and control	Poor waste management and lack of control can mean excess waste is produced on site, and if unlicensed carriers are used there is the risk of fly tipping.
Reducing climate change	Sites use energy in their operations and the production and transport of construction materials. Both will produce carbon as a by-product, which can lead to climate change.

Discussion Point

How could you practically monitor the transport aspect of MAN03?
How could you confirm that all timber was sustainable?

may be used to calculate the miles travelled to and from the site for each visit, and to develop a total of the miles travelled for the project.

To ensure that timber is sustainable, all deliveries of timber must be accompanied by formal certification, for example, from the Forest Stewardship Council (FSC), to prove it has been manufactured from sustainable sources. This will apply to timber in the form of doors, window frames and joists, as well as finished timbers in the roof trusses and skirting boards. It is a good idea for the construction manager to set up a file to log for all timber products coming on to their sites, with copies of the relevant certification for each delivery load.

Discussion Point

How could you find out the previous land use of a site?
Where would you visit for information?

Previous land use can often be identified on old maps of the area, which record the types of industry in the area (e.g. tanneries, dye or chemical works), or show previous features, such as parks or houses. Such maps may be obtained from the local library, and some libraries even have a dedicated local historian. Local residents can also be a good source of information, including knowledge of the activities carried out on the site over the years.

You may also obtain historical information from the internet through various archives, but many of these charge for access.

Discussion Point

How would you prevent the spread of knotweed on the site? What practical measures could you take?

Physical barriers could be used to segregate the work area to ensure that operatives not involved in the removal works could access areas contaminated by knotweed. Boot and wheel washes could also be installed to ensure that no parts of the plants are tracked on to areas as yet uncontaminated by feet or plant wheels and tracks. Education of the workforce is also essential, so that people know why the measures are in place and the consequences of any actions. Communication through training, tool-box talks and posters will help make sure that people appreciate the seriousness of the problem.

Discussion Point

The site office and welfare cabins can also use a lot of energy and water.
How could you reduce consumption in these areas?

Energy use in the cabins can be reduced by selecting low energy-use cabins that have as much natural light provision as possible and energy-efficient heaters. The location of the accommodation should be as sheltered as the site layout will allow, and include simple measures such as fitting closers to external doors to ensure that heat remains contained. Movement sensors should be used to ensure that lights are not left on unnecessarily. Some new cabins even have solar panels fitted to their roofs to provide alternative energy sources. Energy use for the site accommodation should be metered and monitored to make sure that there are no sudden spikes in use which could indicate misuse of the facilities.

Cabin roofs may also be modified to collect rainwater for flushing the toilets and urinals, which reduces the unnecessary waste of drinking water for this activity, and the urinals may be changed to flush on use or even be made waterless in their operation. Taps within welfare facilities should be timed so they cannot be left running.

3.41

Discussion Point

What common construction activities could produce run-off?
How could you plan to contain it?

All activities involving water can be the cause of run-off, for example, using wheel washes at the gate or hosing down roads. Another common construction practice is the washing out of concrete wagons on sites, which again creates polluted water run-off. The EA have produced a Regulatory Position Statement: *Managing concrete wash waters on construction sites: good practice and temporary discharges to ground or to surface waters* (2011), which sets out good practice for their management. On small sites with few concrete deliveries some discharge may be permissible without a permit, but on larger sites other control measures may be required to store, let settle and reuse the wash water where possible. For more details refer to the guidance.

When considering water run-off, the best solution is to plan construction activities to ensure that they take place on a managed hardstanding so that water cannot leech into the ground, and any run-off is collected and managed as necessary. Bunds can direct water flow to ensure that they are channelled into the correct drainage systems or swales for settlement.

3.42

Discussion Point

What would you need to plan for if this material was to be used on your site?
How would you communicate this to the site team?

The correct PPE would need to be to hand, and could be included in the spill kit closest to where the material was being used. The material would have to be kept on a hardstanding or drip tray. A spill kit containing absorbent granules would be needed, rather than cotton pads, and plastic shovels to clear any potential spills. Hazardous waste provision would also be needed.

In order to communicate this, a TBT should be carried out with those who are using the material and working close by. This should supplement the method statement for those working with the material. A dedicated response team should be established and a contact number provided on a sign near the material or spill kit so that the team can be quickly summoned in case of an incident.

Waste can initially be avoided by minimisation through design (see Section 3.7.3), and the selection of materials that are robust and unlikely to be damaged and therefore wasted when they are being stored on site.

Within site management, materials ordering and management is a priority for reducing the production of waste. Materials must be carefully planned and controlled to ensure that only materials needed are brought to site with no over-ordering which can also result in waste. However, misuse of materials within the build can also lead to waste; for example, the use of a high-specification acoustic block where it is not required because the correct blocks are not on site due to under- or late ordering. This is both a waste of money (higher cost) and energy (embedded in the block).

A large amount of waste on site is caused by poor materials handling and management, and more details about how to avoid this with good practice may be found in Section 3.2.3.

Discussion Point

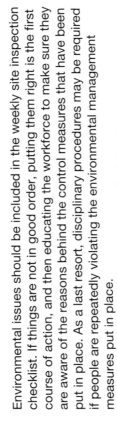

3.44

What other environmental management measures could be considered?

How would you check and control them?

Environmental issues should be included in the weekly site inspection checklist. If things are not in good order, putting them right is the first course of action, and then educating the workforce to make sure they are aware of the reasons behind the control measures that have been put in place. As a last resort, disciplinary procedures may be required if people are repeatedly violating the environmental management measures put in place.

An example checklist could include the following:

Energy and water conservation:

- Are office lights turned off when not in use?
- Are heaters turned off when not in use?
- Is plant turned off when not in use?
- Are any hose connections leaking?
- Are any taps dripping?

Materials management:

- Are the stores clean and tidy?
- Are all materials stacked safely?
- Are all materials free from contamination?
- Are any materials being damaged?
- Are materials being handled efficiently?
- Are materials easily and safely accessible?

Waste management:

- Are the skips identified as necessary?

- Is waste being placed in the correct skip?
- Is hazardous waste segregated?
- Is the skip area clean and tidy?
- Are consignment notes being completed correctly?

Prevention of pollution:

- Are spill kit contents complete?
- Is there clear signage for the spill team, including contact names and numbers?
- Are there drip trays under all fuelling points?
- Are polluting materials stored on drip trays/hardstandings?
- Have those working with polluting materials read their MS?
- Do those working with polluting materials know what to do in case of a spill?

Ecological management:

- Is plant being regularly maintained?
- Is any plant producing fumes?
- Is dust being damped down?
- Is run-off being controlled?
- Is all TPO fencing secure?
- Is all TPO fencing the correct distance from the trees?
- Other measures as required.

Some of the answers to these questions can be confirmed visually, but others need the construction manager to talk to the workforce to check that they are aware of the issues and know how to act should an environmental incident occur. Inspections can be supported by TBTs, briefings and other formal training to ensure that everyone is kept up to date.

3.45

Discussion Point

Why do you think CSR could help profits?

CSR is able to improve the image of a company and therefore attract clients and investors. By association a client can claim a more sustainable and responsible approach to their own activities, if their construction projects are being built by a company with similar values.

Recently there have been several scandals where large manufacturers have had to admit that although their own industry activities are socially responsible, those supplying components or other elements for their products or services are not meeting the same responsible standards; for example, suppliers using child labour in overseas factories to provide components for UK products.

To ensure that CSR is embedded through the supply chain, a company needs to be able to trust and also support good practice in all its suppliers. Although foreign child labour is cheap, it is not socially acceptable for consumers in the UK, and so a company will have to think about how to manage this in terms of potentially increased costs.

3.8 Construction management in practice: summary

Part 3 has provided practical knowledge of how construction managers plan and control the various production parameters in their daily work.

The key tools in production planning include the construction programme, used to plan and control time and money through the allocation of resources. The programme must be supported by effective site management, articulated through the site layout and able to develop and change to ensure that it remains the most efficient layout throughout the project. Good time and site management will support cost planning and control to ensure that realistic tenders are submitted, including those preliminary items needed for site management, and that the work can be tightly controlled to hopefully achieve net gains on the project. These will in turn support high-quality production, allowing the time, environment and costs for the necessary labour, plant and materials to enable the workforce and the construction manager to deliver a defect-free project to the client.

Within these more commercial elements of production management, the health and safety of the workforce should be prioritised to ensure that everyone is able to return home safely every day. The UK construction industry needs to do much more to improve the safety and health of its workforce, and this must be led by construction managers. Environmental management also has an important role to play, and small actions on sites can support the construction industry in its quest to become more sustainable and help reduce man-made climate change on a global scale.

All of these production parameters involve elements of planning and control, and must be considered holistically to ensure effective and efficient construction management in practice.

Part 4 explores some ideas around production management, including approaches which suggest that improvements can be made through collaborative working and the adoption of new technologies such as BIM. These new directions in construction management aim to develop processes that are able to optimise all the parameters of production and always deliver best value to construction industry clients.

Further reading

More details on **site management** may be obtained from the Environment Agency and the HSE.

Advice and guidance from the Environment Agency specifically for the construction industry may be found at www.environment-agency. gov.uk/business/sectors/136246.aspx.

The HSE's guide to organising a safe and healthy construction site may be found at www.hse.gov.uk/construction/safetytopics/ siteorg.htm.

The HSE also provide a detailed approved Code of Practice to support the CDM Regulations which may be downloaded for free from www.hse.gov.uk/pubns/books/1144.htm.

For all legislation, you need to visit www.legislation.gov.uk and search within it for the regulations you need.

The Considerate Constructors website contains details of its scheme, but also guidance on how to join and how to become a considerate constructor: www.ccscheme.org.uk.

The following is good textbook with plenty of images and case studies to show you different site layouts around site establishment and materials management:

Cooke, B. (2011) *Construction Practice*, Blackwell, Oxford.

Source data for establishing activity durations for **time management** may be found in the following text:

Davis Langdon (eds) (2012) *Spon's Architects' and Builders' Price Book*, Taylor and Francis, London (reissued each year to keep pricing information current; however, durations do not change as frequently).

Some good textbooks which explore programming techniques in more detail, and some alternative methods including network analysis and line of balance, are as follows:

Cooke, B. and Williams, P. (2009) *Construction Planning, Programming and Control*, Wiley-Blackwell, Chichester.

Griffith, A. and Watson, P. (2004) *Construction Management: Principles and Practice*, Palgrave Macmillan, Basingstoke.

More detailed approaches to **cost management** may be found in the following texts:

Griffith, A. and Watson, P. (2004) *Construction Management: Principles and Practice*, Palgrave Macmillan, Basingstoke.

Harris, F. and McCaffer, R. (2006) *Modern Construction Management*, 6th edn, Blackwell, Oxford.

Nunnally, S.W. (2006) *Construction Methods and Management*, 7th edn, Prentice Hall, Harlow.

Ranns, R.H.B and Ranns, E.J.M. (2005) *Practical Construction Management*, Taylor & Francis, London.

Pricing for the elements of construction work may be found in the following text:

BCIS (2012) *Comprehensive Building Price Book 2012: Major and Minor Work*, Building Cost Information Service, 29th edn, Poole.

For more details on **quality management,** the ISO standards website may be found at www.iso.org/iso/iso_9000.

All the Building Regulations may be downloaded for free from www. planningportal.gov.uk/buildingregulations/approveddocuments/.

Some good textbooks which discuss quality management systems in more detail are as follows:

Griffith, A. and Watson, P. (2004) *Construction Management: Principles and Practice*, Palgrave Macmillan, Basingstoke.

Harris, F. and McCaffer, R. with Edum-Fotwe, F. (2013) *Modern Construction Management*, 7th edn, Wiley Blackwell, Chichester.

For more information about **health and safety management,** the following two textbooks cover the subject in more detail and will provide a thorough review of the regulations, practices and processes used on site to manage health and safety:

Howarth, T. and Watson, P. (2009) *Construction Safety Management,* Wiley-Blackwell, Chichester, West Sussex.

Hughes, P. and Ferrett, E. (2011) *Introduction to Health and Safety in Construction*, 4th edn, Elsevier Butterworth-Heinemann, Oxford.

More about safety training may be found on the CITB-ConstructionSkills website for the Site Safety Plus Scheme training courses: www.cskills.org/awards/qualifications_and_courses/site_safety/index.aspx.

Or on the CSCS Official Website, which may be found at www.cscs.uk.com.

The HSE also publishes information to support specific regulations, turning regulations into a more readable and practical form for better understanding:

Health and Safety Executive (2007) *Managing Health and Safety in Construction – Construction (Design and Management) Regulations 2007 Approved Code of Practice*, HSE, Norwich.

Health and Safety Executive (2012) *A Guide to the Reporting of Injuries, Diseases and Dangerous Occurrences Regulations 1995*, The Stationery Office, Norwich.

Free download available at: books.hse.gov.uk/hse/public/saleproduct.jsf?catalogueCode=9780717662234.

All health and safety legislation may be found on the government website at www.legislation.gov.uk. Just search the website to find the regulations you are looking for, using the full title.

The Health and Safety Executive (HSE) website has lots of information about the industry, including example risk assessments and guidance on management practices: www.hse.gov.uk/construction/index.htm.

More details about **environmental management**, and specifically sustainability, may be found on the United Nations web page on sustainability: www.un.org/en/sustainablefuture/.

More information about BREEAM, including an introductory video, may be found on the website: www.breeam.org.

For more information about waste management, visit the WRAP construction website at www.wrap.org.uk/category/sector/construction.

The Strategic Forum for Construction's sustainability: water website may be found at www.strategicforum.org.uk/water.shtml.

All documents, guidance and information about and from the Environment Agency may be found on their website: www.environment-agency.gov.uk.

Advice and guidance from the Environment Agency specifically for the construction industry may be found at www.environment-agency.gov.uk/business/sectors/136246.aspx.

Specific documents they have produced that are very relevant to construction environment management include the following:

The Working at Construction and Demolition Sites: PPG6 Pollution Prevention Guidelines. An electronic version may be found at www.environment-agency.gov.uk/static/documents/Business/EA-PPG6_-_03_2012_Final.pdf.

The Regulatory Position Statement – Managing concrete wash waters on construction sites: good practice and temporary discharges to ground or to surface waters is available at www.environment-agency.gov.uk/static/documents/Business/MWRP_RPS_107_Concrete_washwaters.pdf.

The Guiding Principles for Contaminated Land may be found at www.environment-agency.gov.uk/research/planning/121619.aspx.

The Knotweed Code of Practice may be found at www.environment-agency.gov.uk/static/documents/Leisure/Knotweed_CoP.pdf.

The Natural England website, with more information about ecology and wildlife, may be found at www.naturalengland.org.uk.

For all legislation visit www.legislation.gov.uk and search within it for the regulations you need.

Textbooks with more details and information on environmental management and sustainability are as follows:

Griffith, A. and Watson, P. (2004) *Construction Management: Principles and Practice*, Palgrave Macmillan, Basingstoke.

Morton, R. and Ross, A. (2008) *Construction UK – Introduction to the Industry*, 2nd edn, Blackwell, Oxford.

PART 4

New directions in construction management

Introduction

This part of the book explores some of the most recent developments and new directions in construction management. Some have developed from academic research, some from best practice on sites, and others have been implemented by government to try to drive the industry forward.

Construction has often been described as a 'dinosaur' of an industry, still using techniques that were developed hundreds of years ago and showing a reluctance to adopt new technologies, processes and ways of working. One school of thought argues that 'if it ain't broke, don't fix it', but others point to the all too common problems of project delays, cost overruns, poor-quality work, poor health and safety and environmental records, and argue that there is massive room for improvement. Common thinking suggests that 'if you always do what you always did, you will always get what you always got!'

Part 4 will look at the following areas:

- The various drivers for change in the construction industry.
- What barriers to change exist, and how they may be overcome.
- Lean construction thinking and how this is changing management practices.
- Building information modelling (BIM) and how it is changing the way we work.
- The changing role of the construction manager, and new skills that may be needed in the future.

While the ideas and thinking behind these developments can be explored, they are still evolving and changing, as is the way they influence and affect construction management on sites. It is essential for good construction managers to keep up with new developments, to make sure their projects are run as effectively and efficiently as possible. Through industry journals, attending professional body CPD events and following news stories about the industry, you too can keep up to date with new directions in construction management.

For more information and details, several more advanced textbooks, websites and journals have been recommended in Further reading to help you build on the knowledge gained here.

4.1 Drivers for change

Despite common claims that construction is reluctant to move forward, there is evidence of change.

Construction suppliers and product manufacturers are always seeking to develop new technologies that are more effective and efficient in terms of their cost and installation times, to help increase their markets and company turnover. For example, the development of various cladding systems has changed the colour and finish of many residential projects. Sometimes 'brickwork' is no longer even made of bricks, but often of more time and cost efficient brick-coloured tiles fitted to a framing system, as is shown in Figure 4.1.1.

Construction companies are constantly seeking to implement new products and technologies, as well as to improve their own management practices to ensure they are winning work while shortening their construction periods and reducing costs.

But this type of change, while bringing benefits to construction clients in terms of lower prices or delivery times, does not address the bigger problems. These changes often just help improve and develop individual companies, which are implementing changes for their own commercial success. Such isolated developments, while they do bring some benefits to industry and its clients, often still fail to address the wider industry problems – why the industry still delivers projects late, over budget, of poor quality, and with poor health and safety and environmental records.

While some companies have developed integrated **supply chains** to try to manage problems in their overall project delivery, many have not. Companies may also be contracted to work on projects in traditional ways, which reduces the effectiveness of such proactive management practices. There is more about **supply chain management** and how this can benefit construction projects in Section 4.4.3.

As a result the government, the largest construction client in the UK, has been the main driver in bringing about industry change, trying to address these wider issues. Different governments have requested several reviews of the industry while they have been in power, and produced many different reports, and established a large number of industry bodies to try to implement and manage the recommendations made to directly address industry problems.

Figure 4.1.1 Brick-coloured tiles in a framing system

4.1.1 Industry reviews

This section summarises three of the most prominent industry reviews and the reports they produced. These reviews were also supported by, or gave rise to, many other reports and projects, some government driven and some initiated by industry and its collaborative bodies.

Constructing the Team (1994)

The review carried out by Sir Michael Latham focused on the way projects were procured and the existing ways of working in the industry, to explore why there were so many problems.

In his report *Constructing the Team*, published in 1994, Latham stated that the root of many of the problems lay in the contractual nature of the industry and the adversarial ways of working. He found the industry to be highly fragmented, each company, contractor and subcontractor just looking out for themselves, and this was why the industry was unable to deliver quality and value to its clients.

Latham argued for **partnering** and **collaboration** throughout the supply chain, teamwork and the open sharing of information to

4

enable effective and efficient project management and delivery. He made specific recommendations to the industry and the government to bring about change, including better coordination and integration of the design and construction processes and the need for clients to be more aware of their role in the process. As a result of the report, several industry bodies were established to try to bring about the changes Latham proposed, and some of his recommendations were incorporated within the New Engineering Contract and led to the **Construction Act 1996**. But although there was some shift in how projects were procured, contracted and constructed, in the majority construction management practices remained the same.

4.1

Discussion Point

Why do you think the report did not have a significant effect?
What construction industry practices could have influenced this?

Rethinking Construction (1998)

In 1998 another government review was commissioned, this time led by Sir John Egan, former chief executive of Jaguar Cars and BAA. This review focused on industry efficiency and quality, and drew upon Egan's background in manufacturing to provide new perspectives on problems and how to solve them.

Egan published his report *Rethinking Construction* in 1998, arguing that there was a lot of waste within construction processes and many activities were undertaken that did not add value to the end product. He also believed that arguably the most important people in the projects were construction clients, and that their needs were not being met. He proposed a:

radical change in the way we build. We wish to see, within five years, the construction industry deliver its products to its customers in the same way as the best consumer led manufacturing and service industries. To achieve dramatic increases in efficiency and quality that are both possible and necessary we must rethink construction.

Egan suggested that the industry should look to manufacturing to develop the process needed to deliver client satisfaction through a quality product. Like Latham before him, he argued for **integrated teams** and work processes, and believed that partnering throughout the supply chain was essential to meet these goals. He also suggested that construction should use more **standard components** and modern technologies to reduce waste in both time and cost. Given that many construction projects are the same (e.g. houses), Egan believed that they could be produced using manufacturing approaches rather than creating a new bespoke product each time.

Rethinking Construction did have its critics. It was pointed out that buildings are not quite comparable to cars; they are often necessarily bespoke to meet clients' needs, and so cannot be 'componentised'. For example, foundations are by their nature bespoke to a particular building and the ground beneath it.

Never Waste a Good Crisis (2009)

More recently, a review of the industry's progress was led by Andrew Wolstenholme of Balfour Beatty Management, to reflect on industry change to the proposals made by Egan and Latham. Although the industry had been enjoying an economic boom since 1998, a recession had begun and his report sought to continue to drive positive change despite hard times ahead. His report *Never Waste a Good Crisis* was published in 2009, with forewords by both Egan and Latham.

Wolstenholme found that although changes were occurring within industry practices, they fell short of Egan's aspirations from 1998. While safety and profitability in projects had improved, other areas had seen less change, and the industry was still implementing short-term thinking in terms of business models, which had seriously limited supply chain integration and partnering. The report also found a lack of strategic commitment at senior management levels in both industry companies and their clients.

The long-term environmental impacts of the construction industry were also considered, and Wolstenholme identified the need to think of construction projects in terms of their **whole life cycle**. This meant reframing construction projects beyond the construction period and costs, to also include their operation and maintenance once complete.

Discussion Point

Can you identify key themes across all three reviews?
What are the key areas for change?

4.1.2 Government strategy: Construction 2025

The most recent driver for change in the UK is the government's *Industrial Strategy: Government and Industry in Partnership – Construction 2025*. This strategy was published in July 2013 and sets out the goals and processes the government sees as vital to put UK construction at the forefront of global construction in the future.

Construction 2025 sets several headline visions for the industry:

- To reduce construction costs and the whole life costs of build assets by 33 per cent.

- To reduce the overall time for project delivery from inception to completion by 50 per cent.
- To reduce greenhouse gas emissions in the built environment by 50 per cent.
- To reduce the trade gap between imports and exports by 50 per cent.

In simple terms, the government wants the industry to deliver projects faster, cheaper and greener, as well as to develop an export market for products and materials. For such targets to be met, a fundamental change in how the construction industry operates may well be needed, as well as support from construction clients.

Construction 2025 also focuses on the following five key areas and aims for the industry:

- People – to develop a talented and diverse workforce.
- Smart – to be efficient and technologically advanced.
- Sustainable – to lead the world in low carbon and green construction exports.
- Growth – to drive growth across the UK economy.
- Leadership – to be led by the Construction Leadership Council.

Again, these aims will need industry innovation, government and client support, as well as more fundamental changes in construction management processes to be achieved. A new Construction Leadership Council has also been formed to support the delivery of these joint industry and government commitments through various industry bodies.

The Strategy also sets out several drivers for change:

- Improved image of the industry.
- Increased capability in the workforce.
- Clear view of future work opportunities.
- Improvement in client capability and procurement.
- A strong and resilient supply chain.
- Effective research and innovation.

4.3

Discussion Point

How does this strategy fit with the previous reviews?
Do they share ideas for change?

Construction 2025 has met with mixed reviews from industry. Some feel that its targets are too ambitious, and may negatively affect other aspects such as health and safety or environmental management, while others feel that they are too vague and cannot be measured. Others think the 'strategy' is designed to encourage the industry to innovate on its own, rather than commit any government resources or involvement, although it has been welcomed by some as setting clear direction for the industry to modernise and drive change into the future.

4.2 Barriers to change

As *Never Waste a Good Crisis* found, the industry has done well in some areas; its health and safety record has improved considerably and it has met many of its environmental management targets, especially those around recycling and waste management. But in other areas it has not made such significant progress.

Partnering and collaboration, solutions suggested by all three industry reviews introduced above, were found by Wolstenholme to have limited adoption, although there were a few success stories. Some larger organisations have developed their own integrated supply chains; for example, Laing O'Rourke bought several of its subcontract companies and they now operate as an integral part of the Laing O'Rourke organisation, giving the company the internal capability to provide mechanical and electrical services, in-situ concrete and pre-cast concrete for all its projects.

But this kind of integration is limited to the very large contractors with the capacity to bring their supply chain 'in-house'. Most of the industry remains fragmented, and subcontractors and suppliers remain employed through supply chains established on project-by-project contracts.

It has been argued that the fundamental ways in which the industry works have limited the development of partnering and collaborative working.

Some view construction projects as 'one-offs', and it can be hard for companies to see why investment of time and money in the development of supply chain relations would be worthwhile, if at the end of projects supply chains are disbanded and new ones needed for future projects.

Alternatively, companies could take their supply chains with them from project to project. If they are not needed on the next project, they may well be needed on the one after that. Contractors can also have contracts with their supply chain for many projects running at the same time. Good working relationships can stretch beyond the very next project into the longer term as required.

Clients also need to support the industry by regulating the demand for construction projects, and *Construction 2025* does include commitment from the government (the industry's largest client) to:

Develop and refine the pipeline of future work opportunities and make it more useable for all construction businesses.

This means ensuring that demand for construction work is managed and procured to form a regular work flow for industry, so that companies and their partners and supply chains *can* move from project to project together. If the government could achieve this it would be of great benefit to the industry, and help develop fully integrated supply chains and true collaborative working.

The way construction work is commercially managed is also seen as a barrier. All work is contracted, to establish its scope and terms of payment, but this reliance on contracts can create adversarial relationships with a 'them' and 'us' mentality rather than developing collaborative working. As each organisation is out to make a profit, there are often problems of self-interest. Aggressive commercial practices are common, driving supply chain prices down as far as possible to maximise profits for companies higher up the chain. Contracts are often only for individual projects, resulting in a lack of long-term commitment and encouraging companies to opt for quick individual profits over longer term mutual gains.

Short-termism is common, due to the project-based nature of the industry and the struggle to maintain relationships over several projects. But short-termism may also be found in UK governments themselves, which frequently change the strategy and plans of the previous government as soon as they come to power. For example, the 2010 Coalition government cancelled the previous government's Building Schools for the Future project, with negative consequences for the industry in terms of workload and employment.

Criticisms of comparisons to manufacturing are also common, and the industry often sees itself as unique. The fact that the work takes place outside and in-situ on construction sites, and involves highly complicated multi-trade processes, is often compared to manufacturing which takes place in warm and well-lit static factories, where standardised components are put together by machines in the same way every day.

Discussion Point

What aspects of construction have to be different to manufacturing? In what ways could it be made similar?

This argument is often levelled at those trying to introduce new processes and change to the industry. If the industry is unique, then it must have its own unique ways of working, and trying to implement other ideas and work practices simply won't work. But this attitude doesn't support any progress and development; if air traffic controllers hadn't switched to new technologies we wouldn't be able to fly everywhere for our holidays – the old systems just couldn't have coped with the numbers of planes in the sky today. If construction doesn't develop, we will always do things the same way, and the same problems will likely occur. People do like to stay with what they know and are familiar with – change is often a very hard thing to bring about.

In order to break down these barriers, the industry needs to make progress and prove that there are different and better ways of working.

This is one of the aims behind Constructing Excellence's demonstration projects. But overall change has been limited. In 2008, Sir John Egan reviewed how the industry had responded to his 1998 report, and he summed it up as follows:

In summary, I guess if I were giving marks out of 10 after 10 years I'd probably only give the industry about four out of 10, and that's basically for trying.

(Egan 2008)

Not a great result!

But the industry *is* making efforts to change despite the barriers in place, driven forward by various initiatives generated by government, industry bodies, academics, and of course pioneers within the industry itself. The rest of this part of the book will explore some of the more prominent of these initiatives in more detail, and ultimately find out what these new directions in construction management mean for modern construction managers.

4.3 Whole life cycle construction

A fundamental change in the way industry thinks has been the redefinition of 'construction projects', to go beyond design and construction aspects, and to consider the operation, use and maintenance of finished projects as equally important parts of the construction process.

This idea has emerged as a result of several different considerations:

- The focus on industry clients in order to deliver best value and quality products also means establishing how they would use, maintain and ultimately reuse or dispose of them.
- The focus on environmental thinking means consideration of the environment not just during the construction phase, but also the continuing lifetime of projects and how much energy they will use to light, heat and ventilate when in use.
- The focus on value for money not just during the construction phase and on project completion, but also with relation to the costs to run and maintain projects for the rest of their lifetime.

Whole life cycle thinking was included as a key consideration in *Never Waste a Good Crisis*, and has set the new parameters for construction projects. In order to deliver client needs effectively and efficiently, construction managers must think beyond the site and the construction phase of the work, to the design beforehand and the operation afterwards. Construction management has to think outside its own 'site box' in order to be able to contribute to the modern built environment.

Client need
(Demolish,
refurbish or
new build) → Design → Tender and procure → Award contract(s) → Construct → Handover

Figure 4.3.1 Traditional procurement route process

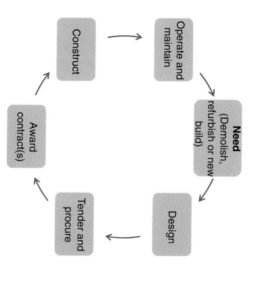

Need
(Demolish,
refurbish or new
build)

Operate and
maintain

Construct

Award
contract(s)

Design

Tender and
procure

Figure 4.3.2 Traditional procurement route cycle

At one time the project was seen as a linear process, as is shown in Figure 4.3.1, which follows the traditional procurement route. This has now become a cycle, as is shown in Figure 4.3.2.

Obviously the sequence of the tender, procurement and contract award stages depend very much on the **procurement routes** clients are using.

But the operation and maintenance of the completed project now has as much value as the design and construction phases. In fact, the capital costs of a construction project, the cost to actually build it, is far smaller than the costs to maintain and operate for the rest of its life once it is completed. Operating costs have been estimated by *Constructing Excellence* (2004) to be between five and ten times as much as capital costs for projects. Clients' needs have consequently become the focus in the cycle, and whether they intend to demolish, refurbish or build new has become the ignition point for the construction process.

This shift to a client-focused industry is what Egan suggested in 1998 and means that clients' other needs have also been included. This has led to a growth in facilities management as a specialist field, in order to provide the skills needed to manage and maintain modern buildings and ensure that they are operating as effectively as possible.

The following new terms have emerged from within construction management:

1. **Buildability** refers to the process of construction, ensuring that the design can be efficiently, effectively and safely built in reality.

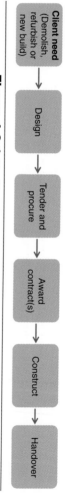

Construction teams should be able to influence and help develop project designs that can be constructed to meet all client needs on site. For example, a change in cladding system may mean that scaffolding is not needed, saving time, money and risk to the workforce.

2. **Usability** refers to the use of projects once completed, ensuring that the end users are able to use the building in the way it was intended by clients; for example, making sure a new factory design meets the needs of the factory foremen and their staff when they come to work inside it.

3. **Maintainability** refers to the ease with which a project can be maintained for the duration of its lifetime. This could mean the inclusion of self-cleaning glass in the windows to save on long-term maintenance costs and eliminate safety risks to maintenance staff who would otherwise have to access the windows to clean and maintain them.

Each one of these characteristics of a project influences the others. Figure 4.3.3 shows the backward effect of these characteristics within the cycle, and therefore the new way in which information must be communicated up the chain to ensure that these elements are fully considered.

This means that as well as more integrated project teams in terms of design and construction, clients must also become more involved in their projects and include their operation and maintenance teams in the process. The whole life cycle of projects has now become the concern of construction managers, and they must ensure that their contributions to construction projects include awareness and knowledge of the cycle as a whole.

4.4 Lean construction

4.4.1 What is lean construction?

Lean construction is a different way of thinking about construction management. It developed in response to the common problems in

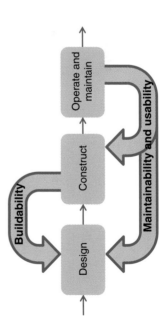

Figure 4.3.3 Backward flow of influence within the whole life cycle

construction projects, as well as under the influence of many industry reviews.

Lean Construction was initially put forward as an idea in the early 1990s. Lean suggested that the ideas of how work was planned and managed, the theories of construction production management, didn't actually match the realities. For example, it was established that only about half of all work was actually completed as planned. This would obviously be a problem in terms of time management, and could easily lead to project delays and overruns. Lean thinking argues that this can be significantly improved, simply by thinking differently about the construction process as a whole.

Lean Construction took ideas from other manufacturing industries, and like Egan, looked to the car industries, especially the Toyota way of working. By thinking more about the *process* of construction and not just focusing on the end *product*, benefits could be achieved.

Lean Construction puts forward the idea of construction as *project-based production*, which means that construction projects become both the process *and* the product.

Through this approach, problems, defects and waste throughout the construction process and in the end product could be identified and removed.

Lean thinking focuses on the need to remove waste from the supply chain, which can take many forms including time, money, resources and effort:

- There may be waste in lengthy tendering and procurement processes, which can be eliminated through the use of partnering and **framework agreements.**

- There may be waste in legal battles and contractual disputes, which can be eliminated through collaborative working.

- There may be waste in materials, which can be eliminated through coordinated design practices among architects, subcontractors and their suppliers to create new products that work together producing minimal waste.

- There may be waste of time due to poor communication processes and lack of information sharing, which can be eliminated through shared work practices and the use of new technology such as BIM.

Removal of such waste would result in best value for clients, who would get the best construction project they could within their specified parameters.

Lean construction thinking applies to the whole project life cycle. At all stages waste can be removed in the form of wasted efforts, wasted time, wasted money, not to mention the physical waste that fills up skips on sites. Every time waste is removed, value is added to the process and end product, making a better project for clients, and better time, cost, quality, health and safety and environmental management for contractors, increasing their profitability and improving their reputation.

4.4.2 Lean thinking

To put lean construction into practice, construction managers need to **think lean**. The project becomes a production system, and construction managers and their teams should try to eliminate waste throughout the process, not just in the end product.

This means that the production **processes**, the way the project will be managed and constructed, should be designed and planned alongside the final product, the project itself. When architects and engineers are creating their designs for buildings, construction teams should also be involved to think about the process of how they will be built. This increases the buildability of the project.

Buildability means that the project is designed to allow for the most effective and efficient construction methods, using standard components and **modern methods of construction** where possible. This also means that the health, safety and environmental impacts of the process of construction can be considered while the project is being designed, and impacts removed and reduced where possible before the project gets to site. Lean thinking reduces waste in terms of time and money, for both the client and the project team.

Lean thinking also requires true partnering and **collaborative working**: project teams and their supply chains become one collective enterprise. They can then work together to remove waste from projects and share the benefits.

For example, contractors may be able to each make 10 per cent profit from a project, but if they worked collaboratively and sought to reduce waste throughout the project process, they might *all* be able to make 12 per cent profit on the project, while also giving the client an overall cost saving of 5 per cent on the originally quoted price. This is collaborative working for mutual project gain – everyone is a winner.

A more detailed example would be a potential change to a large structural panel design that would mean that a cheaper standardised window product could be manufactured and fitted more quickly by the window subcontractor. As a result the panel manufacturer should be financially compensated for the saving in time and money this new panel/window interface design will bring to the project – although it will be the window subcontractor who will gain the immediate benefits. For this to work, information must be freely shared to allow the client and project team to effectively manage 'lean', and time and money across the whole project and across traditional contractual boundaries. This is where it can become complicated, given the nature of traditional construction contracts, but that does not mean it is impossible.

For lean construction to work there must be commitment from the entire project team. There should be strong leadership in all contributing organisations to make sure people buy into lean thinking and work together to achieve the maximum possible reduction in waste and therefore best value for clients.

Lean thinking means that people need to step outside traditional contractual relationships and work collaboratively, share information and develop shared processes for manufacture and construction that bring mutual benefits to clients, project teams and all the way down the supply chains.

4.4.3 Lean working

For construction managers, lean construction thinking can be transferred into several ways of lean working. These include the following:

- **Just in time** production and delivery.
- Lean coordination through **last planner.**
- Modern methods of construction (innovative design and assembly processes, including off-site manufacture, prefabrication and pre-assembly).
- Fully integrated supply chains and supply chain management.
- Use of new technology such as building information modelling (BIM).

The first four of these work practices are introduced here; BIM is examined in more detail in Section 4.5.

Just in time

Just in time (JIT) is a process designed to improve production and reduce waste that was originally developed from manufacturing processes and brings a more detailed level of precision to construction management.

JIT means that the right quantities of the right materials are procured or produced by the supply chain and delivered at precisely the right time to projects for inclusion in the works. This results in cost-effective production using the minimum amount of resources.

For example, on construction projects materials are often 'over-ordered' just in case they run out, but this leads to wasted money in buying unnecessary materials which are vulnerable to damage and theft, and usually end up being wasted. Over-ordering is common because if materials are 'under-ordered' this leads to standing labour, which is expensive, and lost time in production.

But these habits have developed from a lack of confidence in the supply chain. If the right materials *did* arrive precisely when ordered for, then there wouldn't be the need to over-order just in case, as there would be confidence that the right materials would arrive at the right time.

For JIT to work, the whole supply chain must be involved and fully integrated. A supply chain with a proven quality record can be established, making sure that requirements are clear and can be met; this means that work is done right first time, eliminating waste in terms of both defective materials and the time taken to carry out rework. A job role, such as a materials scheduler, is often created to manage the process for main contractors to carefully measure quantities, check specifications and schedule deliveries.

Discussion Point

4.6

Can you link JIT to total quality management?
How could it help develop a company's TQM?

JIT needs people to change their thinking about the construction process, not just the end product. The supply chain should extend to the raw materials suppliers who also need to be involved to make sure they are able to make their contributions effectively and on time for the rest of the supply chain to meet their obligations as well. The supply chain knows precisely what it needs to provide and when, and can plan work accordingly to meet the requirements of lean construction.

Last planner

Last planner also brings more detail to the process of production management. It focuses on the site production phase of the project and is based on the idea that the work should be planned by those who will be involved in carrying it out – the last planners in the chain. This will often be the construction manager, the site foreman and the subcontractor supervisors, rather than the pre-tender planner sat in the company head office.

Using the last planner process, shared plans are made on a weekly basis, taking into account what can *actually* be achieved during the week, based on the knowledge of the people who will be carrying out the work. This reduces variation in the construction process: only what can be achieved is planned for; coordination issues, access, materials delivery and logistics may all be discussed, which results in a precise and accurate weekly plan of the work. Brief catch-up meetings are held each morning to review the plan, note any changes and monitor the work as it progress through the week.

Last planner recognises that there is a difference between what *can* be done and what *should* be done, and the site team are able to plan what *will* be achieved. Although the longer term programme is driven by the milestones in the contract programme, *how* these will be achieved is planned by those who actually carry out the work.

This approach reduces waste by creating a realistic plan that may be used to manage materials through JIT, ensure that labour and plant are optimised, highlight any logistical issues, eliminate bottlenecks or other delays that might emerge, and reduce waste throughout the construction process.

Modern methods of construction

Lean thinking also asks for innovative design and assembly processes. This may include off-site manufacture, standardisation, prefabrication and pre-assembly, all of which are also known as modern methods of construction (MMCs).

The term MMCs can be misleading. The industry has always used some standardised components – for example, bricks are produced to standard and consistent dimensions. They are a standardised component in terms of size (within tolerances); it is colour, strength and other properties that vary. As a consequence, window frames are also (ideally) standard sizes so that they fit easily within brick-coursed openings, and blocks are also standard sizes to enable wall ties to bed easily between the mortar joints of the two leaves.

But lean thinking goes beyond this. Badly managed and executed brickwork on site can lead to various different kinds of waste; for example:

- Waste in materials if bricks are badly stored, cut inaccurately, too easily damaged or simply sunk into the mud and forgotten.
- Waste in quality if mortar drips (known as snots) are not cleaned from the wall ties, creating a moisture bridge across the leaves.
- Waste in time if the weather turns too cold for bricks to be laid.
- Waste in health and safety if scaffold lifts are needed as the work goes higher, adding risks for the bricklayers and the scaffolders.

Lean thinking asks whether there is a way to reduce this waste, saving time and money for the client and the supply chain.

And new ways have developed. Brickwork panels can now be made in a factory environment, brought completed to the project and fitted to the steel or concrete superstructure. This removes the effects of weather and mud from the work, bricks can be better stored and pre-cut as needed, no snots are dropped as the work progresses and panels can be built in bespoke access frames, removing the need for scaffolding.

Innovations in cladding have meant that traditional brickwork is becoming rarer on large new-build construction projects, although it is still popular in domestic construction. On larger projects, prefabricated panels are being used to clad buildings, such as the pre-cast concrete wall panel being fitted below in Figure 4.4.1.

This has saved the need for external scaffold lifts as the panel has been specifically designed to be fitted from inside the handrail of the in-situ concrete frame. The panel is also pre-finished, meaning that once installed it is complete, and no further work has to be done to it, again reducing work at height.

Use of MMCs has meant that many elements of the construction process are now prefabricated and only brought to site to be fitted together. This ranges from mechanical and electrical services

Figure 4.4.1 Modern methods of construction: pre-finished wall panels

constructed in prefabricated sections that only require lifting into place, to entire plant rooms which simply need bolting together and connecting up.

The use of prefabricated rooms, known as pods, has become common in residential and hotel construction, where entire bathrooms are made in the factory – all fittings and fixtures included down to the mirror and soap dish, and simply slotted into the building frame and connected up to the main services.

Such developments are becoming more common as supply chains work more closely together to develop new solutions that remove waste from the construction process.

Supply chain management (SCM)

Construction industry supply chains are often complex; they are very fragmented and will often overlap. Supply chains include all parties and companies that contribute to the construction process and its end products.

Within one element of construction work – for example, the brickwork on a large project – the main contractor may subcontract the work to a specialist subcontractor. The specialist subcontractor will in turn have materials suppliers that provide the bricks and mortar, bricklayers to lay the bricks, and a plant supplier to provide the forklift to load the

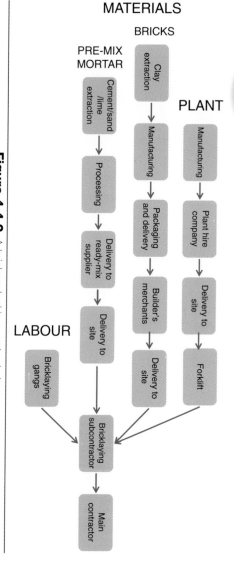

Figure 4.4.2 A brickwork skin supply chain

bricks on to the scaffolding. All of these suppliers will have their own suppliers, manufacturers and processors as well.

This small supply chain for just the brickwork (not including any blockwork, ties or cavity insulation) has already become quite complicated, as is shown in Figure 4.4.2.

Figure 4.4.2 shows the subcontractor's three main supply chains – labour, plant and materials – and all the different companies and parties involved in just the brickwork, and how they all become part of the main contractor's supply chain for the work.

There are often many more companies involved in a supply chain than just those who carry out the works on site.

Discussion Point

What supply chain members are involved in a concrete pour?
What supply chain members help provide a steel frame?

Contractors, and construction managers, rarely think in these terms, as all of these processes are often allocated to just one subcontractor who would then manage the various elements as part of their own supply chain. But construction projects are no longer limited to the work that occurs on site; they are now considered through their whole life cycle, and this also applies to supply chain management.

SCM involves developing relationships throughout the supply chain that support mutual goals and enable the development of significant competitive advantages. This means a change from traditional procurement processes, which are often used to drive down subcontractor and supplier prices to maximise the profits of those higher up the chain. Project-by-project contracts are used to divide work up into different packages, with no one willing to work outside their own remit as there would be no benefit.

Instead of looking for the lowest price in their tender returns, buyers should instead look for best value, quality products and processes, and the potential to develop future relationships and shared innovations. This way of working requires the establishment of frameworks by clients and large contractors to establish long-term relationships with their subcontractors and suppliers to reduce waste when tendering for work. Work can be quickly allocated to framework members, who can work together as an integrated supply chain.

Cooperative forms of contract should be used, with partnering protocols and agreements in place to support and allow collaborative working. Open-book commercial management is needed to make sure everyone is in a win/win situation from the partnership and can see where savings will be made and shared. Supply chains need to be involved in projects from the very start, so that they can contribute their knowledge and skills to develop best value solutions while eliminating waste at the early design stages of projects when the greatest benefits can be achieved.

Such relationships must be developed, nurtured and continually managed to ensure collaborative working. This takes time, resources and commitment from all members of the supply chain, including construction clients. There needs to be leadership buy-in from both construction companies and their clients to develop a non-adversarial culture, and companies must look beyond the next short-term project, and to long-term plans instead.

4.4.4 Benefits of lean

Lean construction enables a better understanding of the client's needs all along the supply chain, enabling full application of lean thinking and working to projects to support best-value delivery. Sir John Egan is a strong supporter of lean construction, and stated that the use of partnering through the supply chain has shown that 50 per cent reductions in costs and 80 per cent reductions in project time are possible in some cases (Egan 1998: 9).

Waste is reduced throughout the construction process, resulting in improvements in the quality of the end product. The use of continuous improvement as part of the management process also means that good practices may be shared across the supply chain and, where things have gone wrong, **lessons learnt** can also be shared to make sure the same mistakes are not made a second time.

The development of supply chains and frameworks can also mean a steadier workload for contractors, subcontractors and suppliers, allowing them to become more secure about their future, reducing the need for cut-throat tendering, and enabling them to start to integrate and innovate within the supply chain. This also means that wasted costs for tendering and re-tendering are removed, there is certainty of payment throughout framework projects, and there are fewer disputes due to the right-first-time approach.

But for these to become long-term benefits for the industry and industry clients as a whole, the savings of time and money need to be passed down supply chains to all members. If benefits are hoarded by those at the top of the chains, then the chains will break and cease to work properly.

4.4.5 Criticisms of lean

Despite the potential benefits of lean, which have been proven on many construction projects, it has not been welcomed with open arms by all parts of the industry.

The traditional barriers to change remain, as discussed in Section 4.2, and the contractual way we organise our work sets out 'them' and 'us' from the very start. Companies seeking to maximise their individual profits, and the project-based nature of the industry do not readily support long-term supply chain management and the development of fully integrated teams.

Others also argue that lean can only be used on large projects by large organisations, and that it is not practical for smaller projects. The development of frameworks within SCM has bundled many smaller projects into one large one; for example, school repairs are no longer tendered on an individual project basis but bundled into one large regional project. This cuts out the smaller contractors who could have won an individual project or two, and limits the work winning to the large contractors who can undertake such large projects. Although these mega-contractors could themselves subcontract out some of the work to the smaller contractors, and some do, the framework approach has been accused of leaving the SMEs out of the process. Some local authorities have sought to address this issue by having different frameworks for different sizes of project, but many SMEs feel that they are not being helped by this change in industry working.

Some even suggest the industry is too set in its ways to change. But these arguments all miss the fundamental point that lean is a difference in thinking to be applied where possible. Lean looks for the win–win approaches in construction management, and is vital to improve industry performance and stop delivery of projects late, over budget, to poor-quality standards, and with bad health, safety and environmental records. Such thinking may be applied to projects of any size and any duration. Lean construction may be seen as simply good practice construction management which employs thoughtful and in-depth production planning and control mechanisms to reduce waste on projects.

Although it may take time for industry to adopt lean thinking and replace traditional ways of working, as clients become more aware, and more involved in projects through whole life cycle considerations, they are more likely to want to build lean and may drive it forward.

You can find out more about lean construction from the Lean Construction Institute UK. This is a charitable organisation that supports research and dissemination of lean construction principles and best practice, aiming to improve construction processes within public sector works.

4.5 Building information modelling

4.5.1 What is building information modelling?

Building information modelling is more commonly referred to as BIM. BIM has been described as *the* management tool to bring about industry change. BIM is able to support the collaboration needed for lean construction, enable integrated supply chain management and working, allow construction projects to build for the whole life cycle, and finally bring the construction industry into the digital age.

But what exactly *is* it?

Even the UK government's BIM Task Group struggles to define BIM, stating that:

There are many definitions of what BIM is and in many ways it depends on your point of view or what you seek to gain from the approach. Sometimes it's easier to say what BIM isn't!

- It's not just 3D CAD
- It's not just a new technology application
- It's not next generation, it's here and now!

BIM is essentially value creating collaboration through the entire life-cycle of an asset, underpinned by the creation, collation and exchange of shared 3D models and intelligent, structured data attached to them.

(BIM Definition 2014)

BIM may therefore be seen as both a management **tool** and a management **process**.

The **tool** takes the form of a computer-generated 3D model, contributed to by all necessary parties, which creates a true virtual representation of the proposed construction project. This is the next step forward from 2D Computer Aided Design (CAD) to designing in 3D. Various computer software packages may be used to develop the model. Specialist designers will often have their own packages that meet their specific designing needs, but they will need to be interoperable with other packages, so there must be early agreement as to the software programs and interfaces to be used. An example BIM model is illustrated in Figure 4.5.1, produced by Adept Consulting Engineers Ltd, showing the foundation and structural design of a project.

Figure 4.5.1 A BIM model

The **process** is the approach made in the development of this 3D model. For the model to be successful and effective, people need to share information and work together. They need to collaborate on areas of the design to reduce waste and optimise value for money, and resolve any problems in the virtual world, rather than when problems occur on site and so are much more expensive to fix.

4.5.2 BIM beyond design

Many people confuse BIM with the 3D model itself, and think of BIM as simply a design tool; but it has capacity for far more than that and can support the management of projects throughout their whole life cycle.

It can include the time management of the project (known as the fourth D; 4D), to sequence the work and build virtually within the real life space, positioning cranes or other access equipment to check the buildability of the design, as well as health and safety during construction.

It can include the costing of the project (known as the fifth D; 5D), including prices for digitised components as they are added to the design, and seeking out the potential to reduce cost and add value, by selecting alternatives from suppliers' virtual product catalogues that are themselves linked to the model.

It can also model the operation of completed buildings within various parameters, predicting energy use and carbon production once buildings are in operation. It is able to advise clients and their facilities management teams of the expected running costs and maintenance schedules for the products used in the construction, revealing the true usability and maintainability of projects.

BIM is integrated with another government initiative for the UK construction industry: **government soft landings** (GSL). Both BIM and GSL turn the construction project life cycle on its head and start the process with a focus on clients' and end users' needs in terms of the use and operation of buildings, making the process 'begin with the end in mind', involving clients and their facilities management teams throughout the construction process. GSLs also intends to make the transition from construction to client ownership much smoother, ensuring that clients and their teams are able to quickly manage their new buildings after handover.

The government's BIM Strategy (2011) also incorporates a US development called COBie, the **Construction Operations Building information exchange**. This is a non-branded database which contains various information sheets documenting projects. These are populated with information about the systems and assets within projects that will need managing, including product information, spares, warranties and maintenance needs.

COBie is able to provide a detailed, client-friendly handover system, replacing the old and often poorly compiled **operation and maintenance (O&M) manuals**. COBie also enables clients to clearly see the design performance of projects, what maintenance needs scheduling and what operational costs are likely to be. COBie data can be easily compiled and extracted from the BIM data as projects progress.

If projects can first be constructed virtually, with all these elements considered, it is easy to see why BIM has become so prominent so quickly in the industry. Building virtually before work starts on site has the potential to solve many problems around time, cost, quality, health and safety and the environment, through improved design, buildability and construction management. BIM is also able to support the drive to improve value and reduce carbon performance across the whole life cycle of the project. Linking through BIM to GSL and COBie also means that construction projects can deliver better client satisfaction with the end product, providing them with projects that not only meet their needs but also have enhanced usability and maintainability.

4.5.3 Where has BIM come from?

BIM is not a new development in the global construction industry; countries other than the UK have been using BIM for some time, and seeing the benefits. Major projects around the world have used BIM, including the Helsinki Music Centre in Finland, and the Sydney Opera House refurbishment project in Australia.

But the UK construction industry has been slower to change. As a result the UK government launched its BIM Strategy in March 2011. Uptake of BIM is to be achieved through a 'push-pull' approach. The push aims to ensure that all players reach a minimum performance in BIM use by 2016, while the pull will come from the clients in their specification of BIM within their projects.

The government is driving BIM with two key elements in mind:

- Whole life cost (improve value for money).
- Carbon performance (reduce outputs and improve sustainability).

These are reflective of the new whole life cycle construction – they are both **end user** considerations but reflect the influence usability and maintainability have on the design and construction processes.

They also demonstrate lean construction thinking. Improved value for money is at the heart of lean construction, supported by reduction in waste, which itself is able to influence improvements in carbon performance.

The government has established a BIM Task Group to help deliver the BIM Strategy. Many industry bodies are involved, including the Construction Industry Council, and the Group is:

> supporting and helping deliver the objectives of the Government Construction Strategy and the requirement to strengthen the public sector's capability in BIM implementation.
>
> (BIM Task Group 2014)

4.5.4 What are the 'levels' of BIM?

In order to enable a measured adoption of BIM in the construction industry, and to clearly explain how BIM may be used and its use increased, the government has set several BIM levels, as is shown in Table 4.5.1.

The government has set a target for all government projects to be 'Level 2' BIM by 2016.

In real terms Level 2 BIM means that all project designs should be produced as 3D models, and there should be a managed environment in which the different designers' 3D models may be shared. This does not mean that there is only one model which everyone works on, a common misunderstanding for BIM at Level 2,

Table 4.5.1 Levels of BIM

BIM level	In real life this means ...
Level 0	Unmanaged CAD probably 2D, with paper (or electronic paper) as the most likely way data and information are shared in the supply chain.
Level 1	Managed CAD in 2 or 3D format using BS 1192:2007 (the collaborative production of architectural, engineering and construction information: Code of Practice) with a collaboration tool providing a **common data environment**, possibly some standard data structures and formats. Commercial data managed by stand-alone finance and cost management packages with no integration.
Level 2	Managed 3D environment held in separate discipline 'BIM' tools with attached data. Managed commercial data. Integration on the basis of proprietary interfaces or bespoke middleware could be regarded as 'pBIM' (proprietary). The approach may utilise 4D time and 5D cost elements.
Level 3	Fully open process and data integration. Managed by a collaborative model server. Could be regarded as iBIM or integrated BIM potentially employing concurrent engineering processes.

Source: UK BIM Strategy (2011).

but rather all designers will have their own models, which may need to interface with each other at set times that are stated in the contract documents.

The BIM Task Group have developed several documents to support the necessary processes to achieve the 2016 target, and various working parties are involved in helping support BIM uptake in industry.

Plans are already under development for helping industry achieve Level 3 BIM after 2016, which will involve more integrated design and one shared model for projects.

4.5.5 BIM terminology and documentation

The BIM levels above introduce some new terms, and make reference to some key documents that help clarify what BIM is about and the changes it makes to industry practices.

At the time of writing, BIM use in industry is still developing towards Level 2. Lessons are being learnt from completed BIM projects, and new projects where BIM is more integrated into the design and construction process are starting to be built.

As an industry we have not yet achieved the lean construction practices of fully open collaboration and information sharing which meet Level 3 BIM. Rather, the industry is taking steps towards the Level 2 goal, which is more restrictive but still much more collaborative than what has been done before. Level 2 BIM shares information at key points in design development, rather than everyone working on just one shared model, yet it still enables better communication and integration of the supply chain through the process.

BIM is all about information, and the government have produced a suite of documents to help standardisation, clarify what is needed and when, and how it is to be shared.

They have produced the following documents:

- Employers' Information Requirements (EIRs)
- The BIM Protocol
- Publicly Available Specification (PAS) 1192:2
- Scope of Services for Information Management.

Employers' information requirements (EIRs)

EIRs define what models need to be produced and when. They contain technical details, such as the type of software to be used and the level of detail required, management details, such as how the work is to be planned and what methods will be used for clash detection, and commercial details, such as the timing of the deliverables in order to fulfil the contract.

EIRs introduce the term **data drops** to refer to the shared delivery of specific BIM models, drawings and other information as planned for at various project stages. For example, a key data drop will be the provision of information needed to go out to tender to appoint main contractors for projects. Other information may also include cost models for designs, COBie data, and information about sequencing or even temporary works or site logistics.

The BIM protocol

This is a supplementary legal agreement which can be bolted on to standard construction contracts through the use of a simple amendment, creating additional obligations and rights for employers and contracting parties.

The protocol also has two Appendices which aim to ensure the adoption of collaborative working. Appendix 1 includes all the project-specific information and information from the EIRs. This becomes the contractual **Information Release Schedule** (IRS) for projects. Appendix 2 contains information about project standards, ways of working, and naming conventions for the models and designs.

The protocol takes legal precedence in any contractual arrangements and intends to make collaborative working and the use of BIM processes express contract terms.

It is changes to contracts and the way the industry has traditionally worked that has raised concerns around BIM. As well as the barriers discussed in Section 4.2, some have felt that BIM raises legal problems of its own.

A key area for legal discussions is that of liability. If all supply chain members are contracted for specific work, liability for a specific design failure can be easily assigned. Shared design through BIM may make it harder to allocate responsibility, and if an error is made worse or causes more errors in later design, then issues of liability become much more complicated. There may also be concerns around commercially sensitive information and copyrights. Ownership of BIM models may also be legally difficult to define, especially when the models will be used for the whole life of projects, and the question has been asked 'how far will liabilities stretch?'

So far, the answer is that 'we don't yet know'. The government's BIM documents, and specifically the BIM Protocol, have been designed to clarify and allay such fears and encourage the adoption of BIM in industry. At the time of writing, the BIM Protocol had not yet been tested in court, but if BIM is really encouraging lean construction and collaborating working, theoretically this should never happen.

PAS 1192:2

This document is the *Specification for information management for the capital/delivery phase of construction projects using building information modelling*. It is the BIM-step forward from BS 1192:2007, the Code of Practice for the collaborative production of architectural, engineering and construction information mentioned in BIM Level 1 above.

PAS 1192:2 has been produced by the government's BIM Task Group to support the principles of collaborative working during the project delivery phase to BIM Level 2 working.

The 'capital/delivery phase' refers to the design and construction phases of projects. Another document, PAS 1192:3, is being developed to cover the maintenance and use of projects to complete the whole life cycle.

The document explains and sets standards for the information delivery process at various stages of the process, detailing how information is to be shared and the steps which parties or designers must take in order to have their information approved and included in the shared model. This process may be simplified, as is shown in Figure 4.5.2.

All parties to the design contribute to the shared model held within the project's common data environment, through 'data drops' as defined in the BIM Protocol. From this shared model authorised construction information is produced. During the construction phase, ownership of the model will pass to the main contractors to enable them to share the information with their own supply chains.

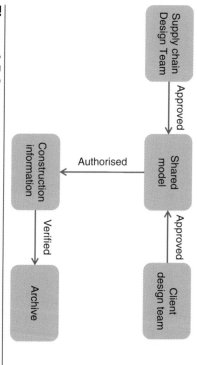

Figure 4.5.2 Flow of information through PAS 1192:2

Following the construction work, this information is verified as a true representation of '**as-built**' and then archived.

PAS 1192:2 notes that for Level 2 BIM, the shared model may actually be more than one model, and will be supported by other information and documents. It is at Level 3 BIM that just one truly shared model will be produced.

Scope of services for the information manager

Another way the government have aimed to support BIM integration is in the creation of a new role in the BIM Protocol, the **information manager**, to deliver information management services. It has been developed from traditional project information management activities, and will normally supplement the work of an existing member of project design teams.

The government have produced a scope of services for the role, which includes establishing and managing the common data environment, undertaking project information management, and facilitating collaborative working, information exchange and management of project teams.

There is no design responsibility in the role; it is a management consideration and part of the BIM process, although if the role is undertaken by designers their own responsibilities for design remain. For example, responsibility for clash detection is a design role and so remains the responsibility of the designers concerned.

All of the BIM Documents may be found on the BIM Task Group website, and are updated as new documents and guidance are produced.

4.5.6 BIM in practice

BIM use is growing within the UK construction industry. While not everyone is operating at BIM Level 2 as yet, use of BIM Level 1 is becoming more commonplace as companies start to share their data in a managed way to improve design processes and collaboration. These earliest applications of BIM at Level 1 involve clash detection and avoidance – spotting where a planned pipe work run clashes with the steel frame, for example – improving design coordination and producing take-offs.

Level 1 BIM also results in the production of models that focus on specific elements of a project, and may then be used by others to develop their own designs and installation proposals.

For example, the design of a housing project produced by Adept Consulting Engineers Ltd may be seen in Figure 4.5.3, but the model also contains details of the work within the ground; the drainage layout may be extracted from the model, as is shown in Figure 4.5.4.

The provision of this drainage layout within the 3D model is helpful to other service providers; it will enable optimum positioning of, for example, the electrical ducts and water main branches within the ground, and ensure that there are no clashes, as well as helping to avoid any unnecessary reworking of excavations.

There are still concerns for small companies, for whom the cost of the software, the new computers to support it and training their staff will be a considerable outlay. Some have argued that this is no different to the technology jump that came in the 1990s when typewriters were replaced by computers, but it is still a big hurdle for SMEs to overcome to become part of a BIM-based industry.

For large companies BIM is becoming the norm. Multidisciplinary consultants such as Arup and Atkins have been offering whole life

Figure 4.5.3 Housing project BIM model

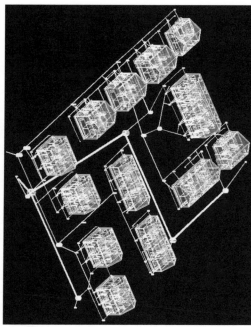

Figure 4.5.4 Drainage layout for housing project BIM model

BIM services for some time, while the contractor Laing O'Rourke has publicly stated that its use of BIM was key in winning them the 'Cheese grater' project in London. The larger organisations are leading the way towards BIM Level 2, with collaborative working, shared data models, applications of BIM for whole life costing, building use simulations, and support and provision of information for facilities management. They are also able to support their supply chains in cascading BIM use down to smaller contractors as they become involved in larger projects.

As clients are becoming more aware of the benefits of BIM, they are starting to follow the government's lead and to make BIM a contractual requirement on their projects.

4.5.7 Benefits of BIM

BIM has been suggested to improve the following aspects:

- Design – reducing on-site clashes, detailing better interfaces and connections between work elements and packages.
- Information management – 80 per cent of those who work with BIM find that it improves the coordination process (NBS 2012).
- Construction periods – time on site is reduced thanks to the elimination of problems in the virtual build.
- Cost savings – 65 per cent of those working with BIM found that it delivered cost efficiencies (NBS 2012).
- Value for money – the client, designers, contractors and suppliers can collaborate to embed value throughout the project.
- Carbon reduction – through enhanced collaboration and simulations of building performance.

BIM is also able to support change in the construction industry for the better. It is able to facilitate lean construction thinking, driving collaborative working, integrated supply chains and improved construction performance through MMC and JIT management.

It is able to deliver the new vision of the construction project, involving clients as key members of supply chains, enabling consideration of the whole life cycle of construction projects from their end use backwards, encouraging enhanced value and reduced carbon along project supply chains.

It is hoped that collaborative working through BIM can support the significant shift needed to change old industry practices, and enable the industry to rethink the way we procure, contract, design, construct and manage our construction projects in the future.

4.6 The changing role of the construction manager

Some of the new directions in construction management included here will not result in significant changes to the role of construction managers. For example, collaborative working has always been a fundamental part of construction management. Getting people to work together in the construction site environment can be challenging and difficult, but it has always been essential to support effective and efficient working, and to ensure that projects are completed within client-specified production parameters.

Discussion Point

What other contemporary issues already form a part of construction management?
How are they currently managed on sites?

It will be more of a challenge to break out from traditional contractual ways of thinking, and develop true collaboration with the supply chain. This will require openness and information sharing, embracing the idea that win/win is better than win/lose every time.

Good construction managers will need to ensure that they keep up to date with developments in construction materials and processes, as more modern methods of construction are adopted in project designs. They must ensure that they are not only aware of new technology, but also the associated requirements of planning and quality that the new processes need, and how they could affect health and safety and environmental management on sites.

Whole life cycle construction thinking will necessarily involve construction managers earlier in projects, to bring their knowledge of buildability to the design and help develop the time and site management aspects of BIM models. Their knowledge of production management will be essential to ensure that the BIM model is itself effective and efficient to construct in reality as well as in the virtual world.

A significant change will be in the use of technology, and new skills will be needed to use BIM in site offices and even on site, employing tablets and handheld computer devices to take the virtual BIM model to the real life project.

Many construction managers already have the IT skills to interrogate online document control systems for construction information, and use 2D CAD drawings to extract design information, details and dimensions. With BIM Levels 2 and 3, these skills will need to be enhanced, to enable the interrogation of 3D models for the same information, as well as to extract time and cost management data needed to coordinate and manage the work on site. Construction managers will need to be able to feed back into the model to control the works and report as-built information as the works progress into the BIM model and COBie.

The role of the construction manager should shift from fire-fighting problems caused by poor design coordination and adversarial working practices, to coordinating and facilitating modern construction projects with enhanced buildability, having helped project teams to plan and eliminate issues around time, cost, quality, health and safety and environmental issues in virtual BIM models.

4.7 New directions in construction management: summary

This part of the book has looked at some of the new directions in construction management.

Change has long been called for by many voices, including construction clients, the government and those working in industry. Problems of project delays, cost overruns, poor-quality work, and poor health and safety and environmental records are evidence that there is room for improvement. New ways of thinking have offered potential solutions, and they are slowly being included as part of standard industry working.

Lean construction is central to the changes, seeking the reduction of waste and the addition of value throughout the design and construction process. Clients have been placed at the heart of construction projects through the shift to whole life construction which emphasises the usability and maintainability of the end products.

Open and effective supply chain management has become essential to ensure that these values cascade down through project subcontractors and suppliers, making the elimination of waste and the addition of value a win/win situation for everyone involved.

BIM has become the tool to drive this change, ensuring collaboration through shared work practices and the shared building model, bringing the industry into the digital age. Through GSL and COBie, the focus has shifted to the production of projects which embody value for money and carbon performance, while providing an effective and efficient built environment asset for clients.

As the title indicates, these are new directions in construction management, but only at the date of publication of this book. Changes will continue to happen and good construction managers must keep up to date with the latest developments, to make sure their projects are as effective and efficient as possible. Through industry journals, attending CIOB and other professional body CPD events and following news stories about the industry, you can keep up to date with new directions in construction management.

Further reading

You will find more information about lean construction and supply chain management in the following texts:

Fryer, B., Egbu, C., Ellis, R. and Gorse, C. (2004) *The Practice of Construction Management*, Blackwell, Oxford.

Harris, F. and McCaffer, R. (2013) *Modern Construction Management*, 7th edn, Wiley-Blackwell, Chichester.

For more information about lean construction and supply chain management, visit the lean construction institute at www. leanconstruction.org.uk/ or the International Group for Lean Construction at iglc.net/.

All the information on BIM, government soft landings and COBie may be found through the BIM Task Group website at www.bimtaskgroup.com. The government's Construction 2025 Strategy can be found at www.gov.uk/government/uploads/system/uploads/attachment_data/file/210099/bis-13-955-construction-2025-industrial-strategy.pdf.

Keep an eye on the industry press: *Building* magazine has been following the ups and downs of BIM as it becomes more common industry practice, and there are often articles in the Chartered Institute of Building's *Construction Manager* magazine and *Construction News* that report on new practices. You can also download the Construction Index App to keep right up to date with announcements and developments in the industry as they happen.

Discussion point comments

Discussion Point

4.1

Why do you think the report did not have a significant effect?
What construction industry practices could have influenced this?

The report asked for a fundamental change in the way construction work was organised. The project-based nature of the industry and the use of short-term contracts on a project-by-project basis are not a good fit with collaboration and partnering in the longer term. The industry would have had to fundamentally change how it procured work, and the benefits of partnering and supply chain integration had not yet been proven to a level that convinced the construction industry to adopt these new ways of working.

Discussion Point

4.2

Can you identify key themes across all three reviews?
What are the key areas for change?

All three reports stated that the construction industry needed to integrate its ways of working, and believed that collaboration and working together was the best way to move the industry forward. Latham promoted partnering as the way to achieve this, and a shift away from traditional contractual practices. Egan also called for integration as well as innovation of new technologies and standardisation through shared product developments. Wolstenholme also focused on the lack of integration at the time of his report, and a lack of partnering in practice.

A key problem appears to be the contractual nature of the industry, how work has been parcelled up, and so individuals are reluctant to step outside these lines to collaborate and innovate together, for fear of negative commercial repercussions. It could be argued that the short-term vision of the industry is in part to blame for the lack of long-term relationships, but there are also issues of trust that need to be resolved as well.

Discussion Point

4.3

How does this strategy fit with the previous reviews?
Do they share ideas for change?

Construction 2025 appears to focus more on the industry outputs rather than the processes of getting there. While the previous reviews promoted collaboration, integration and partnering as necessary for the production of such outputs, Construction 2025 seems to focus more on the outputs themselves. Although there is recourse to the need for future work planning, supply chains and innovation, it is less clear about the potentially more fundamental changes that need to be made in terms of procurement, contracts and litigation.

You can read all the original documents, they can be found online, and see whether you agree with the government's new strategy: do you think it will help give the industry a new direction?

Discussion Point

4.4

What aspects of construction have to be different to manufacturing?
In what ways could it be made similar?

Manufacturing thinking aims to streamline the process, which could be used within construction to develop regular supply chains – suppliers that are able to contribute to every project and so work together to produce the best-value construction on site. Although many elements of construction do need to be designed, many construction materials and components may be standardised to meet certain design parameters – and some already are, such as engineering and facing bricks.

Some have suggested that because construction projects must be placed in their final location they cannot be produced in a factory, but this is limiting thinking to the whole project – the elements that make it up could easily be factory produced, and again, many already are.

Another argument is that projects are bespoke to the client – each one is different. They should be treated as unique, although there are many elements to construction – floors, ceilings, cladding – which could be standardised. For example, there is no need to redesign the connections between the cladding and the frame every time; these could be developed by the cladding and frame designers to find the best solution and so become standardised. Some feel that with standardisation comes a lack of design flair, but there is arguably scope for this within the finishes of a project, keeping standardisation to the structure – very much like some cars.

Another key argument is that there are sudden large shifts in demand for construction products and this affects the relationships and the way the industry works. Some construction projects are more similar to manufactured products which are often marketed for sale (e.g. housing developments), but others could not be built on speculation to then sell. Infrastructure projects need clients to initiate them and so with no clients the industry does grind to a halt.

Discussion Point

4.5

What other examples of materials waste on site can you think of?
How would JIT help prevent them from happening?

Two examples would be as follows:

- Over-ordering of materials can also lead to wasted materials, as they are likely to get damaged while they are being stored on site. Large stockpiles of materials are also vulnerable to theft and may need to be moved around the site (double-handled) so that they are not in the way, which can also lead to damage to the materials. The weather can also cause deterioration to many materials. JIT would mean that the right materials arrive precisely when they are needed.

- Waste can also be caused by using the wrong materials in part of the work, because the right ones are not on site; for example, using a more expensive acoustic block where it is not actually needed, simply because there are no standard blocks on site. This is a waste of materials and money.

Discussion Point

4.6

Can you link JIT to total quality management?
How could it help develop a company's TQM?

It has been estimated that 50 per cent of an organisation's quality issues are related to defective purchased materials. JIT seeks to establish a supply chain with a proven quality record and so will help support TQM practices. This needs supply chain management and full supply chain integration to ensure that the quality requirements are clear and can be met. Work will then be done right first time, eliminating waste in terms of both defective materials and the time taken to carry out rework.

Discussion Point

4.7

What supply chain members are involved in a concrete pour?
What supply chain members help provide a steel frame?

A concrete pour supply chain could look like Figure 4.4.3, while a steel frame supply chain could look like Figure 4.4.4.

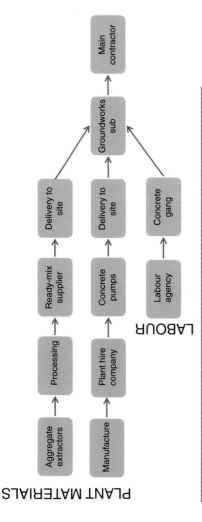

Figure 4.4.3 A concrete pour supply chain

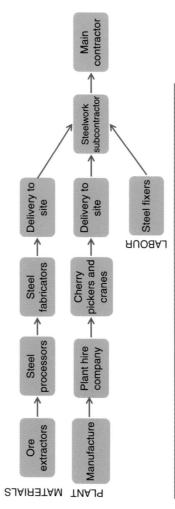

Figure 4.4.4 A steel frame supply chain

Discussion Point

What other contemporary issues already form a part of construction management?
How are they currently managed on sites?

Two examples include the following:

The construction manager must already manage complicated supply chains on site, as different subcontractors and subcontractors-of-subcontractors come to work on projects. They have to ensure that subcontractors have ordered the right materials and have the right labour and plant to complete their works as planned. They need to make sure that all the subcontractor's supply chains comply with the health and safety and environmental requirements of the project, as well as deliver to the necessary time, cost and quality standards. The construction manager sometimes has to force supply chains to comply, but where collaborative working has been developed everyone is working together to achieve a win-win situation.

Construction managers already aim to reduce waste on their sites in terms of materials management and control. Through good site management, plans should have been made for deliveries and storage of materials to minimise the potential for waste and damage, while reducing the possibility of wasted time if materials are not on site. The change to JIT will mean reducing these stores, but construction managers will still need to draw on their planning and management skills to make sure the right materials arrive at the right times.

Introduction to Construction Management: summary

This textbook has hopefully provided you with an introductory insight into construction management.

The complexity of modern construction, coupled with the need to meet the key production parameters of time, cost, quality, health and safety and the environment while managing people, plant and materials on site, can make modern construction management a highly complex process.

The construction industry as a whole has been explored, to provide the backdrop for construction management, and the details of how industry professionals and trades come together in various ways to deliver projects to construction clients has been established.

The theories behind management have been outlined through Fayol's management functions and principles, as well as the need for good people management through teamwork, motivation and leadership. Good construction managers need to be able to lead their projects to success, and depend on construction's people to make that happen.

Construction management practice relates directly to the parameters of production. Through practical examples the processes of production management – time, cost, quality, health and safety, environmental and site management, as well as the ways construction managers both plan and control these interrelated aspects of their work – have been examined in detail.

New directions in construction management have also been outlined. Although some may not seem very 'new' they are still not common practice in the UK construction industry, but good construction managers can bring the ideas of lean construction, supply chain management and BIM to their projects and help develop the industry of the future.

Construction management is challenging and often hard work, but it can also be highly rewarding. Construction managers need to be well qualified, people friendly and knowledgeable about their industry, keeping themselves up to date, using their professional bodies as a vehicle to do so. With changes in technology and improvements in collaborative ways of working helping to spark new innovations, it is an exciting time to be involved in the construction industry.

Sources

BAM Construct UK Ltd (2011) Sustainability Policy, BAM, available at http://www.bam.co.uk/brochure/BAM_401_csr_display.pdf [15 November 2013].

BCIS Comprehensive Building Price Book (2013) *Major Works*, 30th edn, Section 6, pp. 2–3.

BIM Strategy (2011) UK Government BIM Strategy, available at http://www.bimtaskgroup.org/wp-content/uploads/2012/03/BIS-BIM-strategy-Report.pdf [22nd October 2013].

BIM Definition (2014) BIM Task Group FAQs, available at http://www.bimtaskgroup.org/bim-faqs/ [12 January 2014].

BIM Task Group (2014) About Us, available at http://www.bimtaskgroup.org/about/ [12 January 2014].

BIS (Department for Business, Innovation and Skills) (2013) Construction 2025 Strategy, available at https://www.gov.uk/government/uploads/system/uploads/attachment_data/file/210099/bis-13-955-construction-2025-industrial-strategy.pdf [5 December 2013].

Bostik (2012) Safety Data Sheet: Cementone Black Waterproofing Paint, Bostik, available at http://www.bostik.co.uk/support/safetyData/diy/product/245/44 [2 April 2014].

BREEAM (2012) BREEAM New Construction Non-Domestic Buildings, Technical Manual SD5073-3.4:2011, BRE Global Limited, available at http://www.breeam.org/BREEAM2011SchemeDocument/#_frontmatter/coverimage.htm%3FTocPath%3D_1 [14 November 2013].

Calvert, R.E., Bailey, G. and Coles, D. (1995) *Introduction to Building Management*, 6th edn, Butterworth Heinemann, Oxford.

CCS (2013) Site Registration Monitor's Checklist, Considerate Constructors Scheme, available at http://www.ccscheme.org.uk/images/stories/ccs-ltd-section/downloads/2013/2014-Checklist.pdf [29 May 2014].

CIC (2013) tweet @CICtweet 26 November 2013, 13:26 from Westminster, London #talk2013.

CIOB (2010) An Inclusive Definition of Construction Management, The Chartered Institute of Building, available at http://www.ciob.org.uk/sites/ciob.org.uk/files/Redefining%20CM%20(low%20res).pdf [11 June 2013].

CIOB (2013) About the CIOB, available at http://www.ciob.org/about [8 December 2013].

CITB (2013a) Construct Your Future campaign, CITB, available at http://www.citb.co.uk/news-events/campaigns/construct-your-future/ [8 December 2013].

CITB (2013b) Our Role in the Construction Industry, available at http://www.citb.co.uk/about-us/who-we-are/our-role-construction-industry/ [9 December 2013].

Constructing Excellence (2004) Whole Life Costing, Constructing Excellence, available at http://www.constructingexcellence.org.uk/pdf/fact_sheet/wholelife.pdf [18 February 2014].

Cooke, B. and Williams, P. (2009) *Construction Planning, Programming and Control*, 3rd edn, Wiley-Blackwell, Chichester.

Department for Communities and Local Government (2012) National Planning Policy Framework, available at https://www.gov.uk/government/uploads/system/uploads/attachment_data/file/6077/2116950.pdf [16 November 2013].

EC (2013) Small and Medium Sized Enterprises – what is an SME? European Commission's Directorate General for Enterprise and Industry, available at http://ec.europa.eu/enterprise/policies/sme/facts-figures-analysis/sme-definition/ [5 December 2013].

Egan, J. (1998) *Rethinking Construction*, Department of Trade and Industry, HMSO.

Egan, J. (2008) 'Egan: I'd give construction about four out of 10', Building Research Establishment Limited, available at http://www.bre.co.uk/filelibrary/pdf/CLIP/SirJohnEgan21-05-08.pdf [20th October 2013].

Environment Agency (2010) *The Guiding Principles for Contaminated Land*, available at http://www.environment-agency.gov.uk/research/planning/121619.aspx [18 November 2013].

Environment Agency (2011) Regulatory Position Statement – Managing concrete wash waters on construction sites: good practice and temporary discharges to ground or to surface waters, environment agency, available at http://www.environment-agency.gov.uk/static/documents/Business/MWRP_RPS_107_Concrete_washwaters.pdf [18 November 2013].

Environment Agency (2013) The Working at Construction and Demolition Sites: PPG6 Pollution Prevention Guidelines, available online at http://www.environment-agency.gov.uk/static/documents/Business/EA-PPG6_-_03_2012_Final.pdf.

Environment Agency (2014) Construction, The Environment Agency, available at http://www.environment-agency.gov.uk/business/sectors/136246.aspx [18 February 2014].

Fryer, B., Egbu, C., Ellis, R. and Gorse, C. (2004) *The Practice of Construction Management*, Blackwell, Oxford.

Griffith, A. and Watson, P. (2004) *Construction Management – Principles and Practice*, Palgrave Macmillan, Basingstoke.

Harris, F. and McCaffer, R. with Edum-Fotwe, F. (2013) *Modern Construction Management*, 7th edn, Wiley-Blackwell, Chichester.

HM Government (2013) Industrial Strategy – government and industry in partnership: Construction 2025, available at https://www.gov.uk/government/uploads/system/uploads/attachment_data/file/210099/bis-13-955-construction-2025-industrial-strategy.pdf [20th October 2013].

HSE (2009) 'Company Fined £135,000 Following Death at Isle of Dogs Construction Site', Health and Safety Executive, available at http://www.hse.gov.uk/press/2009/coilon050509.htm [6 February 2013].

HSE (2012a) 'Twenty Year Trends in Worker Fatalities', Health and Safety Executive, available at http://www.hse.gov.uk/statistics/industry/construction/index.htm [19 January 2013].

HSE (2012b) 'Five Steps to Risk Assessment', Health and Safety Executive, available at http://www.hse.gov.uk/risk/fivesteps.htm [14 February 2013].

HSE (2013a) 'Safer Sites – Targeted Inspection Initiative September 2013', available at http://www.hse.gov.uk/construction/campaigns/safersites/index.htm [8 December 2013].

HSE (2013b) 'Work Related Stress', Health and Safety Executive, available at http://www.hse.gov.uk/construction/healthtopics/stress.htm [14 February 2013].

HSE (2014a) 'How Many Toilets Should a Workplace Have?', Health and Safety Executive, available at http://www.hse.gov.uk/contact/faqs/toilets.htm [18 February 2014].

HSE (2014b) Construction Industry, Health and Safety Executive, available at http://www.hse.gov.uk/statistics/industry/construction/construction.pdf [20 January 2014].

ICE (2013) 'About the ICE', available at http://www.ice.org.uk/About-ICE [8 December 2013].

Koskela, L. (1992) 'Application of the New Production Philosophy to Construction', Technical Report #72, Centre for Integrated Facility Engineering, Department of Civil Engineering, Stanford University, CA.

Laing O'Rourke (2013) 'Environment', Laing O'Rourke, available at http://www.laingorourke.com/responsibility/environment.aspx [15 November 2013].

Latham, M. (1994) Constructing the Team, Department of the Environment and the Construction Industry Board, London.

Lend Lease (2009) 'Policy Statement Environment – Group, Lend Lease', available at http://www.lendlease.com/Worldwide/Sustainability/environment-health-and-safety~/media/Group/Lend%20Lease%20Website/Worldwide/Documents/About-Us/Company%20Policies/Environment%20Policy.ashx [15 November 2013].

Maxey, A. (2011) 'Evaluation of the Construction (Design and Management) Regulations 2007', paper for HSE CONIAC, available at http://www.hse.gov.uk/aboutus/meetings/iacs/coniac/130711/m2-2011-2.pdf [19 February 2013].

McMeeken, R. (2010) 'After Kieron', Building, 1 October.

Mintzberg, H. (1976) 'The Manager's Job: Folklore and Fact', Building Technology and Management, Vol. 14, Issue 1, pp. 6–13.

NBS (2012) National Building Specification Survey, available at http://www.bdonline.co.uk/bim-adoption-rocketing-survey-finds/503158.article [12 January 2014].

NHBC (2011) 'A Consistent Approach to Finishes', National House-Building Council, available at http://www.nhbc.co.uk/NHBCpublications/LiteratureLibrary/Technical/filedownload,15912,en.pdf [26 January 2014].

NPPF (2012) National Planning Policy Framework, Government Department for Communities and Local Government, available at https://www.gov.uk/government/uploads/system/uploads/attachment_data/file/6077/2116950.pdf [29 May 2014].

ONS (2012) Construction Statistics No 13 – 2012, Office for National Statistics, available at http://www.ons.gov.uk/ons/rel/construction/construction-statistics/no--13--2012-edition/art-construction-statistics-annual--2012.html [4 May 2013].

Ordnance Survey (2014) Benchmark Locator, available at http://www.ordnancesurvey.co.uk/benchmarks/ [25 February 2014].

RIBA (2013) About Us, available at http://www.architecture.com/TheRIBA/AboutUs/AbouttheRIBA.aspx c [8 December 2013].

RICS (2013) History and Mandate, available at http://www.rics.org/uk/about-rics/who-we-are/history-and-mandate/ [8 December 2013].

Seddon (2013) Our Commitment, Seddon Construction, available at http://www.seddon.co.uk/csr/our-commitment [18 November 2013].

Scientific American (2008) Cement from CO_2: A Concrete Cure for Global Warming?, available at http://www.scientificamerican.com/article/cement-from-carbon-dioxide/ [18 February 2014].

Smith, K. (2004) 'Failing Foul?', Construction Manager, 5 May.

Strategic Forum for Construction (2012) 'How to Save Water on your Construction Site', Strategic Forum for Construction, available at http://www.strategicforum.org.uk/HowToBrochure.pdf [18 November 2013].

United Nations Economic Commission for Europe (2005) 'Sustainable Development – Concept and Action', available at http://www.unece.org/oes/nutshell/2004-2005/focus_sustainable_development.html [4 November 2013].

Wolstenholme, A. (2009) *Never Waste a Good Crisis*, Constructing Excellence, available at http://www.constructingexcellence.org.uk/pdf/Wolstenholme_Report_Oct_2009.pdf [20 October 2013].

Glossary of key terms

Activity
An activity is an individual task within the construction programme to be completed as part of the work. The level of detail of the activity depends on the level of programme; for example, 'substructure' could be an activity on the pre-tender programme, but this may be broken down into smaller activities: 'excavation, shutter, reinforcement, concrete' could be activities on the operational programme, but they both ultimately mean the same thing. The level of detail of the activity relates to how the programme will be used; a more detailed list is helpful in allocating resources, and so is more useful to construction managers.

Apprenticeship
An apprenticeship involves people working, training and learning on the job, as well as attending college for academic support, so that students can earn while they learn.

Architect
The architect is the designer of the project, and produces the drawings and specifications for the work. In the UK, they must be registered with the Architects' Registration Board to practise.

As-built
The record of the actual construction, as opposed to the planned and designed construction, which may have changed in practice due to unforeseen issues such as interface problems or design clashes.

Back-to-back
A term which means that all key dates and critical content of a document should match up with any other versions or associated documents. For example, a short-term construction programme issued to the bricklayers should show the same watertight date as the main construction programme; they are 'back-to-back' to ensure that everyone is working to the same plan.

Baseline
A programme's baseline is its original version. The baseline shows the original plan against which the works on site are monitored.

Bidding
Contractors are said to 'bid' for work when they submit their tenders to clients. Bidding is the process of preparing the tender to produce an accurate yet competitive price for a project.

Bills of Quantities (BoQs)   BoQs are used to price projects. They list quantities of materials to be used which can then be priced by estimators to cover the cost of labour, plant and materials. BoQs are used to decide how much money a contactor should be paid in monthly valuations, and are a useful cost control tool.

BRE Green Guide   A reference guide which provides an environmental impact assessment for various construction materials across their entire life cycle.

Brownfield sites   Sites on which there has been previous construction or industrial activity. This may mean that contaminants remain from this previous use or other considerations, such as foundations or services within the ground.

Budget   The price for either an element of work or the whole construction project. Historically, many construction projects have gone over budget, and this has given the industry a bad name.

Buildability   The process of ensuring that a design can be efficiently, effectively and safely built in reality leads to the enhanced buildability of a project.

Building Control   The department of local councils charged with monitoring and enforcement of the Building Regulations. Building Control Officers may be involved in approving the design of a project to make sure it complies with the regulations, and also visit the site to inspect the works as they are built. They have the power to stop work if it is not compliant and demand it be removed and rebuilt to meet the regulations.

Building Cost Information Service (BCIS)   A service administered by the Royal Institution of Chartered Surveyors that provides inter alia cost data of completed projects as a basis to predict cost on future projects with similar specifications.

Building information modelling (BIM)   This is a project management tool employed in the process of creating and using digital models for design, construction and operational processes.

Building Regulations   The legislative tool for construction quality standards. The Regulations set out the minimum requirements and specifications for construction work. They are communicated through Approved Documents and enforced by Building Control.

Building Research Establishment Environmental Assessment Method (BREEAM)   A certification system for the environmental performance of buildings, established by a comprehensive review of the design, construction and operation of a project carried out by approved BREEAM Assessors.

Built environment

Refers to everything that is not part of nature; it includes everything man-made that has been constructed, including homes, buildings and infrastructure.

Carbon calculator

A tool for calculating the carbon footprint of construction projects or complete developments. It assesses the materials to be used, the transport costs for deliveries and labour to construct projects, and the amount of energy and fuel used in construction. Added together, these elements form the carbon footprint.

Carbon footprint

The total amount of CO_2 emissions caused by a company in their operations, or produced by its sites during the construction of projects. This may be calculated using a carbon calculator.

Cash flow

A reconciliation of cash/money coming into an organisation from clients versus money going out to the supply chain. Hopefully contractors will have positive cash flow; bank balances in surplus rather than overdrafts.

Cherry picker

A mobile access unit with a small basket on a long, very manoeuvrable mechanical arm. The basket holds the operatives and enables them to work at height and gain access to otherwise inaccessible areas. Working at Heights Regulations apply to their use.

Civil engineering

The profession associated with the design and management of projects such as roads, rail, ports, tunnelling or other types of infrastructure.

Cladding

Any system which provides a building envelope. Cladding can be made of bricks, glazing or large plastic panels.

Clean-up notice

A clean-up notice may be issued by a construction manager to a subcontractor to demand they clean and/or tidy their work area and materials either immediately or within 24 hours. Failure to comply may result in the main contractor carrying out the work, and charging the subcontractors for the time and effort involved.

Client

The person, company or organisation who initiates a construction project and who is paying for it once it is completed. Clients may range from individual private investors to large government departments. Clients often have a team of construction professionals to support them during the project. They may also be known as the employer.

Collaboration

The act of working together to achieve shared goals with mutual benefits.

Collaborative working

This simply means working together. Collaborative working in construction means sharing ideas, being open about how the work is being done and how much it is costing, so that savings may be identified and best practice shared to the benefit of everyone involved in the project.

Common data environment   A BIM term referring to a single source of information for a project, used to gather, collect and pass on all relevant approved project documents for multidisciplinary teams in a managed process.

Communication   The giving and receiving of information. For communication to be effective it must be both received and understood.

Construction Industry Training Board   The CITB is the construction industry's own training board. The CITB sets the standards for trades, provides training and also manages the Construction Skills Certification Scheme.

Construction management   Construction management may be defined as making sure that people are delivering what they should be delivering, on time, to the required quality, within the set budget, to the necessary health, safety and environmental standards. Management must be efficient and effective to meet all of these requirements.

Construction Operations Building information exchange (COBie)   Structured information for the commissioning, operating and maintenance of projects in a neutral spreadsheet format that is used to provide data to clients or facilities management teams to help them develop facilities management and maintenance systems.

Construction Phase Health and Safety Plan   This plan is required by the Construction (Design and Management) Regulations 2007. A plan must be in place before work starts on all notifiable projects. It should contain key information as to how the works will be managed with reference to health and safety; for example, the key risks on the project, how they will be managed, how health and safety information will be communicated to the workforce, and the responsible people in the management team.

Construction Skills Certification Scheme (CSCS)   The CSCS was developed to ensure a minimum level of health and safety knowledge for all construction industry employees. It involves an online health and safety test which when passed results in the award of a CSCS card. These cards are linked to the individuals' work experience and training through the National Vocational Qualification system for their trade, or the supervisory or management level the individual has achieved. Many sites will not allow anyone to work without a current CSCS card.

Contaminated land   Land that has been polluted and now contains chemicals or materials that could become harmful to health and local ecology if they are not correctly managed during construction activities. Contaminated land should be identified at the feasibility stage of a project because if remediation is required it can prove costly and make proposed projects uneconomical.

Continuity of work   Construction managers should always be looking for continuity of work. This simply means making sure that work is available for the labour and plant on projects. It is inefficient if operatives and materials have to leave sites and return,

Continuous improvement   as this can result in delays, additional costs and changes in operatives, and thus variations in the quality of work. Ideally, the construction manager wants to establish their workforce on site and keep them there until the work is complete.

The process of reviewing and evaluating activities so that the next time they are undertaken they can be improved upon. This then becomes a management cycle, and continuous improvement may be applied to all processes to seek out ways to make them better next time round. Continuous improvement should be measurable; for example, projects should seek to improve their waste reduction by an achievable target, project after project, to ultimately eliminate waste permanently.

Continuous professional development   CPD relates to professional body membership and the need for chartered professionals to demonstrate that they are continuing to learn and keep up to date with the latest industry developments. CPD is often carried out by attendance at lectures and workshops hosted by professional bodies or structured reading on relevant industry topics, or by undertaking formal postgraduate courses.

Contract   An agreement intended to be enforceable by law. For construction projects this is usually a legally binding arrangement between two parties to undertake specified work for payment. Among other things contracts detail the work to be done, the date by which it must be completed, the price and the rates of pay for any work over and above that specified in the contract.

Contractors   Any company that works with others through the use of contracts. Often the term is used to show the relationships between companies; for example, the main contractor is in charge of the project, and allocates parts of the work to subcontractors who work for them.

Corporate   Senior management within a company. The Chief Executive Officer (CEO) and Chief Operating Officer (COO) are part of the corporate company team.

Corporate social responsibility   CSR is the thinking that companies and organisations now have a responsibility to the community in which they do their work and to society at large. This often results in social and community programmes being organised alongside construction projects; for example, a school trip around the site or employment of some apprentices from the local area both count as contributing to a company's corporate social responsibility.

Cost control   A system of comparing income with expenditure to identify potential areas of gain or loss, and thus permit timely corrective action to be taken if necessary.

Cost forecast   A prediction of likely future cost to be incurred in the execution of a clearly defined scope of work.

Cost indices Used to measure cost increases or decreases incurred by contractors as the first tier in the supply chain; labour, plant, material and preliminary costs.

Cost planning A system of predicting future likely costs on projects so that clients can ensure that they have the money available to pay contractors for constructing a project.

Critical activities Activities within the construction programme which have no float. They must be completed as programmed, immediately the preceding activity has been completed, or they will delay the end date of programmes.

Critical path This follows the critical activities through the programme. It is usually highlighted in red and focuses the attention of the construction manager on the areas of work that must be closely managed so as not to delay projects.

Curtain walling Any large glazing system used to clad a building and form its envelope. This may be clear or opaque and is held in place by a framing system.

Data drops The transfer of 3D data models and other information as contractually required by the BIM Protocol, in the formats set out in the Employer's Information Requirements which will specify the data needed and the format in which it should be applied.

Datum A quality benchmark level for construction work.

Daywork Work that is executed on the basis of an agreed rate of pay per hour, and is not constrained by a time limit. A contractor may ask an operative to do some cleaning work on site at an agreed hourly rate of £10 per hour, not knowing whether the task may take four or eight hours. All work on site for some operatives or tradespeople may be completed on daywork; thus a construction manager may employ a labourer with responsibility for cleaning, unloading materials from wagons, supporting tradespeople transporting materials on site, etc., at a fixed rate of £10 per hour for 39 hours per week = a weekly wage of £390.

Defects liability period A period of time, often six or twelve months after practical completion, within which the contractor is required to return to site to correct any defects that arise as a result of materials or workmanship not in accordance with the contract.

Demolition The knocking down of part or all of an existing structure.

De-motivators Factors which stop people wanting to work to the best of their ability. They can be physical (for example, a dirty or unsafe working environment), financial (for example, a poor wage for the work being done), or mental (for example, no praise or recognition for a job well done). De-motivators can have a big effect on a construction site team, and the construction manager should try to remove or reduce them where possible.

Design and build (D&B) procurement

A procurement route where the design and construction of a project are bundled together and managed by a main contractor.

Design brief

The initial outline of a project developed by the design team and client to articulate their plans for the project. This may specify the project use, the size of building and the budget for the project, among other things.

Design for demolition

The process of designing a construction project or parts of it so that when the structure is no longer needed it can be carefully demolished and taken apart, and the components used again in a new project or recycled in some other way.

Design team

The design team includes anyone who has design input into the project. The design team usually includes architects, engineers, services designers and other design specialists; for example, acoustic designers would be part of the design team for a new concert hall.

Desktop studies

These form the first stage in a site investigation. They should contain all the information about a site that can be found out without leaving a desk – this includes previous land use, risk of flooding, mining activity in the area, utilities locations, and even previous investigations of the site.

Dilapidation survey

A survey carried out to show the existing state of buildings, roads or pavements near a construction site before any work begins. These surveys may then be used as a record and may be needed to prove that damage was not caused by the construction work, but was already present before work began.

Double handling

This is when materials are repeatedly moved after they have been delivered to site. Ideally materials should be moved once – from the delivery vehicle to the worksite – but often they need to be placed in stores. However, moving materials around once they are in the stores is wasteful of time and can damage them.

Drip trays

Large plastic or metal trays which capture leaks and prevent pollutants from reaching the ground.

Duration

The length of time an activity will take with the specified allocated resources. Durations can be calculated or estimated for all construction activities and indicates the time they will take within the construction programme.

Dust suppression

The practice of using water sprays to keep dust moist so that it does not rise up in clouds. This is useful on sites to prevent pollution of the local area from dust clouds, but if it is ineffective it can also be a waste of water resources.

Dutch auction

A process often instigated by contractors in their relationships with their supply chains, whereby they invite a second round of bidding in an attempt to achieve lower prices than those included in their bids to clients. It is an attempt by contractors to secure net gains. Dutch auctions may also be used by clients in the private sector to achieve lower prices from contractors, but this is much less common. Round 1 bids are sometimes called a 'bid for a bid', and not priced as competitively by the supply chain as round 2 bids, a 'bid for an order'.

Ecological survey A survey carried out by professional consultants to establish what plants and animals may be found on and around a proposed construction site. A full ecological survey will have to be carried out over several seasons to capture the activity of different animals, and during days and nights for nocturnal species.

Embodied energy The amount of energy contained within specific materials that has been used to produce them. This energy remains trapped within them during the lifetime of the structure, and cannot be reclaimed until demolition.

Enabling works These occur at the very beginning of a project. They include establishing the site, setting up the hoarding and access requirements, and may also involve activities like topsoil strip, the construction of temporary roads or hardstandings, or the diversion of existing services. Enabling works allow the main construction works to start on the site.

Envelope The outside 'skin' of a building is referred to as its envelope. For a house this includes the bricks and windows and roof. For a large office block this includes the curtain walling or rain-screen cladding and the roof. The envelope is fixed to the superstructure to make the building watertight.

Environment Agency (EA) The body tasked with the protection and improvement of the environment, and the promotion of sustainable development. It also ensures compliance with and enforcement of environmental legislation.

Environmental The environment is the surrounding area to a project, and includes both the natural and social world. In the natural world animals and plants may be at risk by construction works and so plans must be in place to protect them, for example, to ensure that no pollutants are discharged into local waterways which could kill native species. In the social world, dust is a common side effect of construction work and can negatively affect the air quality in the surrounding area, which may cause breathing problems for people, and so the dust must be controlled.

Environmental Impact Assessment (EIA) An EIA is legally required for some projects and is the process of carrying out a complete assessment of the impacts a project will have on the environment. This assessment involves considering other proposals and justifying why this particular one has been selected, and

Environmental Management Plan (EMP)

explaining what mitigation measures will be put in place to reduce the impact of the project on the environment. The EIA is summarised in the environmental statement for the project.

The EMP sets out the environmental management activities, systems and controls that will be used on the project. This should include how specific environmental issues are to be managed, such as noise or dust, and what steps will be taken to reduce them throughout the project.

Environmental Management System (EMS)

A system through which a company can ensure that it is meeting its environmental commitments and legal obligations. It will include details of all legal obligations and company policy and rules, and how these should be implemented in practice. This will include the identification of those responsible for the implementation of the EMS, who will need what type of training to support the EMS, and how the EMS will be communicated. It may also include the development of a set of environmental documents.

Environmental policy

A company's statement of how it will address and manage its interactions with the environment. This may include aims and targets, as well as references to sustainability.

Environmental statement (ES)

The summary of the EIA. It must be submitted as part of the planning application in cases where an EIA is a legal requirement.

Extension of time

This is granted by a client to a main contractor or a main contractor to a subcontractor, to allow extra time to that originally agreed in the contract, to complete work. The reason for the extension will usually be due to something outside the control of the lower tier party; perhaps exceptionally inclement weather or ground conditions that are worse than envisaged.

Facilities management

The profession which manages built environment assets once they have been completed and are in operation. For buildings, this includes the maintenance and repair of the building fabric and systems as well as its day-to-day running.

Feasibility

The assessment of whether a project is viable to go ahead. This should consider the overall costs of the project, including how much the management of environmental aspects will cost, compared to the potential outcomes in terms of profits or use of the completed project. For example, it may cost too much to decontaminate a particular site, and it may be cheaper to build elsewhere.

Feedback

The process of gathering information for comparison. In construction management, feedback is most likely to involve the collection of information through monitoring about the progress of work on site, and comparing this actual work to the work that had been planned for completion at that moment in time.

Float
Any spare time for an activity to be completed within the programme. Float means that the activity does not need to be completed immediately, but it will lose its float with any delay in starting or finishing the activity, and it can eventually become critical when all float has been used up.

Foreman
Someone with a level of management responsibility for the work, often the person in charge of a work gang. Foremen may carry out some work themselves while also conducting its supervision.

Framework agreements
Long-term procurement agreements by which clients establish a group of contractors, subcontractors and suppliers who agree to provide specific services at a specific price. A framework is not a contractual arrangement, and there is no obligation for work to be allocated, but the idea is that when it is allocated this will be done quickly and efficiently to reduce waste in the tender process. There is also the potential for the framework members to work collaboratively and support continuous improvement on the framework projects.

Gang
Construction teams are often broken down into 'gangs'. These are self-managed units within a company's workforce who often work together. For example, bricklayers often work in a 2+1 gang – two bricklayers and one labourer to mix the mortar and carry the bricks. These gang sizes are used to calculate work outputs.

General operative
They may either be unskilled or have skills in a number of tasks, such as snagging works or basic concreting, and is able to carry out whatever tasks or roles are needed at the time to help the project progress.

Government Soft Landings
The government project, implemented alongside BIM, to ensure the efficient and comprehensive handover of a completed construction project. Also reinforces the whole life cycle of a project to ensure that consideration throughout design and construction is given to carbon reduction and embedded value from clients' perspective.

Green Deal
A government initiative for home owners who will be able to recoup their spend on retrofit improvements to their homes through reduced energy bills. A loan is given to make improvements and this is repaid through the energy bills.

Ground investigations
These aim to establish the composition and types of ground on the site. They involve taking soil samples for chemical testing for contaminants, digging trial pits to examine the layers immediately below the surface, drilling boreholes to explore much deeper into the ground, or carrying out load-bearing capacity tests. The findings of a ground investigation are often used to produce a geotechnical report on the site.

Groundwater
This refers to water that is held in the ground; it is added to by rain and runoff which seeps through the ground to watercourses such as rivers and streams, and eventually reaches the sea. Groundwater may be found at different levels in the ground, which can be determined by trial pits or boreholes.

Groundworkers
These workers are involved with all things in the ground, which includes the foundations and external works such as paving and drainage for a project; this is also known as the substructure which refers to all work below the damp-proof course.

Hardstanding
This can be constructed from layers of compacted stone, or existing site features such as an old car park or playground. Hardstanding is a surface different to the open ground, which can become dusty or muddy depending on the weather. Hardstandings may be used to keep machinery stable, such as cherry pickers or MEWPs, or to provide clean materials storage lay-down areas.

Hazard
A hazard is something with the potential to cause harm. This could be hazardous materials, an unprotected edge or an excavator – there are many different hazards on a construction site, but it is the risk of the hazard that is important. Hazard identification is a key aspect of the risk assessment process.

Hazardous waste
This poses a threat to human health or to the environment. It must be carefully managed, segregated from other waste and only disposed of by licensed hazardous waste carriers. On site, hazardous waste should be stored in a secured skip or drum to keep it separate and to stop water getting into it and potentially causing pollution.

Health
The freedom from illness and disease. Many activities on construction sites can affect workers' health, such as the use of cement which can cause dermatitis, or dust which can cause lung problems. Stress and general well-being are also health concerns that should be considered.

Health and Safety Executive
The **HSE** is the national independent watchdog for work-related health and safety. It has the power to inspect sites and stop work; it investigates accidents, and also provides guidance about the best ways to manage health and safety and comply with the law.

Hedgerows
Lines of trees and shrubs that create a 'wall' of plants, often forming boundaries to fields or land. They contain a variety of species as well as many animals and insects, and are protected in the UK.

Hierarchy of risk control
This sets out risk management in a number of steps, which should be attempted in order to reduce hazards and risks. The acronym ESCAPE sets out the hierarchy – eliminate,

substitute, control, alter work method, PPE and educate – with the best management option first to eliminate the risk or hazard where possible.

Hoardings

These are made of plywood, or other solid sheet material, fixed to posts concreted into the ground, and form the most robust type of site fencing. They are semi-permanent for the duration of the project and are not able to be easily moved. They can provide a surface for advertisement or community projects as well as company information, and may also contain viewing panels so that people can watch the construction work as it progresses.

Hospital jobs

Activities with considerable float which can be dropped on to when other, more critical activities cannot be carried out due to lack of materials, plant or poor weather. They provide work for labour in these cases, and may also involve housekeeping works such as maintenance of walkways or hardstandings.

Hot works

Any construction work that creates heat or could generate a spark.

Housekeeping

The site term for keeping things neat and tidy. Poor housekeeping can lead to poor quality of work, demotivation of the workforce as well as accidents. Slips, trips and falls on site are very common, and are often due to poor housekeeping where materials have not been stored correctly or waste has not been taken to skips straight away. Housekeeping can be managed using clean-up notices.

Housing stock

The term used to describe all the existing houses in the UK.

Information manager

A BIM term referring to the new role assigned under the BIM Protocol to allocate responsibility for management of the common data environment and facilitating collaborative working.

Information release schedule

Details of the necessary information required to complete a project, and the necessary dates for its production so as not to delay subsequent design or construction processes; the IRS can be a contractual requirement for a project.

Infrastructure

The term used to refer to construction projects such as roads, railways and electricity grids. Infrastructure is what supports the buildings and homes in which we live.

Inspections

Inspections form an everyday part of construction managers' duties. They are how managers establish confirmation that something has been done to the correct standards by visiting and physically checking the area of work on the site. There are many different types of inspections; for example, quality inspections may require a visual inspection or physical test of part of the work to check that it meets the specified quality, while part of a health and safety inspection might check that there are no trip hazards or obstructed walkways on the site. Records of inspections often form part of the quality assurance procedures for the site and should be kept safely in the site office.

Integrated teams These are made up of different professions, for example, the architect and engineer, as well as the contractor and specialist subcontractors, who are all working together towards the project goals. Information is shared and work is collaborative to seek mutual benefits for all those involved.

Interim payments A payment made by a client, usually monthly, based on the interim valuation. The payment supports the contractor, and allows it to make payments to its supply chain, including weekly paid wages and salaries, without having to go excessively overdrawn at the bank.

Interim valuations An agreement, usually monthly, between a client's quantity surveyor and a contractor's quantity surveyor about the value of work completed. The interim valuations form the basis for an interim payment. The word interim indicates that it is a valuation of work of a partially completed project.

Invasive species Plants or animals that have been introduced to the UK, i.e. they are not found in the UK naturally. They are termed invasive because the way they grow or reproduce often damages or restricts native species from growing naturally themselves. For example, giant hogweed blankets the ground so that no other plants can grow near it.

Invoice A document raised (paper or electronically) by a party in the supply chain, often at the end of the month, and delivered to a higher tier supplier, as a request for a payment based on work completed or materials supplied.

ISO 9000 standards The internationally recognised quality management and quality assurance standards used by all global industries that have been developed by the International Organisation for Standardisation. Companies can achieve quality assurance by complying with the ISO 9001 standard and passing an independent audit.

Just in time (JIT) A term which means that deliveries of the right materials arrive on site just as they are needed for inclusion in the construction work. This approach avoids the need for large materials stores on sites, where they may become damaged or stolen, and any double handling. Good supply chain management is needed to manage JIT, as close working with suppliers and subcontractors is essential. JIT results in cost-effective production using the minimum amount of resources.

Labour This refers to the people needed to do the work. This is a generic term which includes tradesmen, gangs and operatives, and is used to describe the manpower needed to complete a task.

Labourer An operative without trade skills, but who works alongside trades in a supportive role, for example, by providing the materials needed for the trades to keep working at a steady pace.

Lag The amount of time on the programme after completion of the previous activity that the current activity needs to complete itself.

Landfill taxes Fees that must be paid on any waste sent to landfill. There are different scales for payment, but these are likely to increase over time as the waste hierarchy seeks to reduce the amount of waste disposed of in this way.

Last Planner The final level of operational planning carried out by those on site on a weekly basis. Last Planner software provides the structure to enable this process, and ensures that the detailed planning is followed by those who also carry out the work.

Lay-down area A designated area on site for the unloading and storage of materials too large to be stored in containers, for example, structural steelwork. Lay-down areas should be stoned up where possible and short timbers, known as skids, used so that materials are not resting directly on the ground.

Lead The amount of time the previous activity must have been completed before the current activity can start.

Leadership Leaders set goals for the future and inspire others to work towards them. Leadership is the ability to motivate others and to encourage people to work towards shared goals.

Lead time The length of time between when an order is placed and when the materials or plant can be delivered to the site. For example, roof trusses have to be designed and made to order and so may have a lead time of a month from when the order was placed and when the trusses arrive on site. Other materials can have much longer lead times; for example, steelwork is also made to order but the production time is much longer, and is likely to take several months.

Lean construction Production management focused on the product and process of construction, aiming to minimise waste and maximise value to the client.

Lessons learnt The ways problems have been overcome on previous projects may be used to inform the next project, and stop the problem happening again. For example, the last-minute re-design of a window fixing on site for one project may be incorporated into the design of another, removing the problem and showing that the lessons of the previous project have been learnt.

Level Refers to the setting-out requirements for construction work, specifically the vertical setting out of the work.

Line *Refers to the setting-out requirements for construction work, specifically the horizontal setting out of the work.*

Liquidated and ascertained damages (LADs) *An amount of money paid by the contractor to the client in the event that the project overruns its scheduled completion date as the result of a contractor failing, such as poor management of the project on site.*

Liquidation *The process by which the life of a company (or part of a company) is brought to an end, and the assets and property of the company redistributed. Often arises from a circumstance where a company does not have sufficient funds to pay its debtors.*

Logic *The relationships an activity has with other activities, including sequence and time-scales. Logic is shown on the construction programme using arrows.*

Main contractor *The company that is contracted to the client and takes charge of the construction management of the project. The main contractor often subcontracts elements of the work to specialist companies, although they may also carry out some elements of the project work themselves.*

Maintainability *The ease with which a project can be maintained for the duration of its lifetime.*

Maintenance *This is also a specialist activity that falls within the construction industry remit; for example, repainting domestic properties or replacing roof tiles when needed are both maintenance activities that will be carried out by people working in the construction industry.*

Management *This can most easily be defined as the organisation and control of work, and more complicatedly be defined through the management functions of forecasting, planning, organising, coordinating, motivating, controlling and communicating.*

Materials *This refers to the products needed to do the work. For example, to construct a solid brick wall, bricks and mortar (which could be made up of sand, lime, cement and water) would be the materials needed to complete the task.*

Materials suppliers *The companies who supply the materials used on site in the construction processes, such as the concrete, the mortar, the nails and the sealants.*

Method statement (MS) *A method statement sets out the plan for carrying out work tasks or a package of work (for example, a subcontractor work package, such as flooring or steelwork). The method statement should evaluate the work step-by-step to draw out the hazards which may then be used in a risk assessment. The method statement should also contain details of the management processes that will be used to ensure that the plan is followed, and actions in case of emergency. The written method statement is often used as the plan for a safe system of work.*

Milestones
Key dates within the programme that must be met; for example, the date the envelope is complete and the project becomes watertight, allowing the internal finishes to begin.

Mobile Elevated Working Platform (MEWP)
These are rectangular platforms that can move vertically up or down. The platform is relatively large depending on the size of the machine, and may be used to fit curtain walling and windows. Working at Height Regulations apply to their use.

Modern methods of construction (MMCs)
These may include off-site manufacture, standardisation, prefabrication and pre-assembly, or any new technological development in materials or how buildings are constructed. It can be something of a false label; for example, timber frame is considered a modern method of construction, but timber construction is centuries old; what is modern is the use of prefabricated walls and floors.

Monitoring
Monitoring can take many forms on site. For example, it is how managers control production, generate feedback against the programme, and check that work is going according to plan. Monitoring may be undertaken in an informal way, for example, through simple observations made during a site walk-round or talking to operatives while they are at work; more formal observations can be made by marking up drawings to show work completed, taking progress photographs or completing reports. From this monitoring, the actual work that has been carried out may be checked against the work that was planned, including checking quality aspects of the work, and any discrepancies addressed. Monitoring can also be undertaken of health, safety and environmental compliance, again through inspections and reports.

Motivation
The reason or reasons behind someone's actions or behaviours – this can be either a carrot or a stick. Carrots are incentives to complete work, the most common being a financial bonus, whereas sticks work in the opposite way, and are punishments for not completing work, such as reduced or 'docked' pay for slow or poor-quality work.

Native species
Plants or animals that are found in the UK naturally, and form part of our ecosystem. There is usually a balance between them so that they all have an equal chance of growing and reproducing – together, they make up our ecosystem.

Natural England
The body responsible for the management of the local environment, and provision of guidance, information and licences to manage protected species in England. It is available for consultation on proposed construction projects, and can advise about any potential impacts on the local ecology and what measures should be taken to mitigate them.

Net gains
Companies initially price or estimate for projects net, and then make a judgement about how much money to add, if any, for profit. The profit may be considered as a separate 'pot' of money. Cost reconciliations make judgements about cost

	performance against net estimates. If actual costs are less than predicted estimates, perhaps as a result of efficiencies, a net gain will result. Companies are always seeking to make net gains (extra profit).
Net losses	*Similar to net gains, except that if actual costs exceed predicted estimates, a net loss will result.*
Network analysis	*A method of planning which involves creating a visual network of the activities, durations and sequence. The network is then analysed to show earliest and latest start and finish dates, as well as the critical path and float for the project. Some computer programmes will enable you to plan using network analysis as well as the Gantt chart method.*
New build	*A project that starts on either a brownfield or greenfield site and is built from the foundations up, as opposed to a project that refurbishes or reuses part of an existing structure.*
Nominated subcontractors	*Clients (perhaps on the advice of one of their design team) will nominate a specific subcontractor to complete an element or package of work. Sometimes used for complex trades that need regular maintenance over the life cycle of a building (e.g. mechanical, electrical or lift installations, or where a client is a regular user of a subcontractor on other projects). Less popular currently than it has been in the past, due to legal complexities.*
Notifiable project	*Under the Construction (Design and Management) Regulations 2007, a notifiable project is one which will take 30 days or 500 person days to complete.*
Novation	*The process of transferring a professional from one contractual arrangement to another; for example, the transfer of an architect from a contract with the client to a contract with the main contractor, from the pre-tender to the construction phases of design and build procurement.*
Operation and maintenance (O&M) manuals	*The provision of information to the client on how a building operates and needs to be maintained, often broken down into work packages from a construction perspective rather than the finished product from an operational perspective. O&M manuals are often the cause of client dissatisfaction with project handovers, and are being replaced by COBie through BIM.*
Operational	*This term is used to describe all the activities associated with production on site. Operations for the construction industry are the actual practices of constructing, whether this is building, modifying, refurbishing or demolishing. This distinction is made to separate management – which may be carried out at head office – from operations – which happen on the site. For example, the operational team are the people who actually work on the site every day, and manage the work as it happens. This operational team are also called the site team, and the term operational is often associated with the site.*

Operational planning This is carried out after pre-contract planning and looks in detail at how the work will be done on the site. Operational planning is usually more detailed in terms of activities than earlier planning, as the aim is to produce a programme that may be used by the construction manager on the site. Activities are often broken down by trade or subcontractor to make the allocation of resources easier. Operational planning should clearly identify the critical path and any float in the programme to also help the construction manager as the works progress.

Operational team This is made up of the people who practically build the work, including trades such as bricklayers, joiners and electricians.

Operatives This term is used to describe the individuals who make up the site workforce. It is used for people who do not have any management or supervisory responsibilities; operatives make up the majority of the workforce. Operatives may be skilled craftsmen or general labourers.

Partial handovers These are similar to sectional completions. This means that the project is not totally complete when, for example, a building or part of a building is handed over to the client for their occupation and use. This can mean more complicated management of the site; for example, partial handover of a number of houses in a new estate would mean families moving in while construction works are ongoing close by. This means that extra care and protection of the site is needed to avoid any disturbance or risk to those who have moved in.

Partnering This is the establishment of collaborative working practices built on best value and not just lowest cost. For example, selection criteria could include quality of work or good health and safety performance, as well as value for money. Partnering uses open relationships and the sharing of information, rather than adversarial working practices.

Passivhaus This is the name of an alternative sustainable house design from Germany, which focuses on the building fabric with high thermal performance, airtightness, insulation and mechanical ventilation.

Permits Written permission to carry out an activity. Environmental permits are needed for many different activities and may be issued by various official bodies. For example, to discharge water from the site you will need a permit from the Environment Agency. Carrying out activities without permits may result in prosecution, fines and even jail.

Photovoltaic panels Large panels that use solar energy to produce electricity. They may be fitted to façades and roofs to provide sustainable energy for a development.

Piecework
Incentive schemes where tradespeople or operatives are paid on the volume of work they complete. Very popular with both employers and trades, and consequently very common in construction work.

Planning conditions
Actions that must be met by clients, developers or contractors for projects before they will be granted planning permission. Planning conditions are often applied to planning applications, and state any requirements for the project that must be addressed as noted. For outline planning applications, conditions may need to be met in the full planning application when it is made. They can relate to many different aspects of the project, from environmental concerns to approval of materials and design.

Planning permission
This is required for any construction project of a certain size or type. It must be granted by the local authority's planning office following submission of drawings and other details about the project. Planning permission only lasts for a certain length of time (often three years) and so there can be a rush to start on site before it runs out.

Plant
Construction plant may be either mechanical or non-mechanical. Mechanical plant refers to the machinery needed to carry out construction work; this includes large machinery such as excavators and cranes, to smaller items such as compressors and pumps. Non-mechanical plant has no moving parts and includes scaffolding.

Plumb
This refers to the setting out of a building, specifically its vertically and straightness.

Policy
The term used to describe the government's proposed course of action or principles used to manage a situation. For example, the government has a housing policy which will affect the construction industry with regard to new-build properties.

Pollutant linkage
This is the relationship between a contaminant in the land that could cause harm, the receptor which is something that could be harmed by the contaminant, and the pathway which is the route by which the contaminant travels.

Pre-contract planning
This is carried out after the tender has been awarded and before the contract is signed for the work. The contractor should develop their pre-tender plan so that they are able to confirm more details to the client about the overall production plan for the project.

Preliminary items
Also known as project overheads. All items required on site that are not directly associated with labour, plant and materials for measured works. Main contractors' staff salaries are a major portion of the total cost of preliminary items. It also includes items such as cabins, temporary fencing, temporary roads and temporary power supplies.

Presumption in favour of sustainable development

The statement within the National Planning Policy Framework (NPPF) that planning approval for construction projects that are sustainable should be given without delay. The three dimensions of sustainable development – economic, social and environmental – should be considered, with reference to the local development plan, to assess whether a proposed project is sustainable. If it is sustainable, it should be approved as a matter of course.

Pre-tender planning

The planning of a construction project carried out before the tender price is submitted. Pre-tender planning helps to ensure that the tender price, and methods of work proposed to deliver the project at this price, are accurate and realistic, and also meet the client's specified considerations of time, quality, health and safety, and environmental management as well as cost.

Price

An agreed lump sum to be paid by a higher tier party to a lower tier party in exchange for the completion of a clearly defined package of work. Clients may agree a price with contractors to complete some work for a fixed price. Contractors may agree a price with subcontractors or tradespeople to complete some work for a fixed price, or agree with suppliers to supply material for a fixed price.

Principal contractor

Under the Construction (Design and Management) Regulations 2007, on notifiable projects a principal contractor must be appointed. This is usually the main contractor on the site. The principal contractor has responsibilities under the regulations, such as the provision of a site induction and establishment of site rules, and they must also prepare the construction phase health and safety plan.

Private sector work

Construction work funded by private investment; for example, the construction of a new block of flats to be sold for profit.

Procurement route

The way in which clients buy buildings or other projects from the construction industry. The route used will depend on how the clients wish to share the risks between themselves and other parties (e.g. weather and ground conditions), and the extent to which the design-and-build elements are separated or integrated.

Product

Construction products are those developed from raw materials through a manufacturing process, such as bricks, timber, windows or roof tiles.

Production

The process of making things; construction is a production-focused industry. We produce things – houses, offices, schools, roads and bridges – and so production management is a key part of construction management.

Professionals

The people and companies who design and manage the work, such as architects, engineers and construction managers.

Professional bodies   *A professional body is a recognised organisation that represents a specific profession; often they have been granted a royal charter which enables their members to use the designation 'Chartered', and means that they have to comply with certain professional rules of conduct.*

Professional team   *This is made up of the people who design and manage the work, including members of the design team, such as the architect and engineers, and the main contractor.*

Profit   *A sum of money that is a return on an investment. The difference between money received on a project and total money paid out.*

Programme   *This is the most common way management plans are communicated. A programme is a chart which sets out all the different activities along a timeline, as well as showing how they all link together. The programme is used to control the work, enabling the manager to see what activities should be being carried out at any given time, as well as what activities need to be carried out beforehand in preparation. The most common form of programme is the GANTT chart.*

Programme position   *A comparison of the actual work completed to date against the planned work position. Work can be on programme, behind or ahead of programme.*

Progress report   *A formal report on the progress of work on the site which may be required by head office on a weekly or monthly basis. Such reports can also form part of a company's quality management system. The report should give the programme position of the project, and state how each element of the work is progressing, noting any issues, delays or problems. The report can be accompanied by progress photographs and a marked-up programme to show precisely which activities are on, ahead or behind programme. Contractors will be required to submit monthly progress reports to clients and their design teams.*

Project   *A construction project is initiated by a client for construction work within the boundaries of one financial budget. The term is often used to describe just one construction site, as a self-contained project where the work is focused. However, some projects actually have multiple sites and locations for the work; for example, a project to build several fire stations would actually have several different sites.*

Project duration   *The overall length of time it will take to complete the project. This can be calculated through planning all the activities in the programme, their durations and the logic between them.*

Project success   *This is often judged to be whether a project is completed on time, to the required quality, within budget, and to the set*

Protected species

Plants or animals that have been protected by law. It is an offence to disturb or kill them and if they do need to be relocated permits will be required.

Provisional sum

A sum of money allocated for elements of work that are known to be required, but at the time of putting tender documents together, sufficient information is not available to allow bidders for projects to estimate likely cost. In tender documents, for example, the client will say to bidders: include the provisional sum of £5000 for directional signs.

Public sector work

Construction work funded by the government; for example, the construction of a new school or repairs to roads.

Quality

This can be defined as something being fit for purpose. It refers to the standards set for the work through drawings and specifications. Quality is a key concern within the construction industry, and it is the role of the construction manager to ensure that the work has been carried out using the correct materials and processes to meet the quality requirements such as physical strength, fire resistance or operational ability.

Quality management system

A set of formal processes and procedures used to manage quality within a company. This may include setting out standard processes for establishing quality benchmarks, standard checklists to be used for certain work elements and the timetabling of regular inspections.

Quantity surveyor

The professional role that is charged with the financial management of a construction project. This includes the tender pricing, cash flow management and valuations of the work as the project progresses, and the setting of the project final account. All large construction companies and client organisations will have quantity surveyors working for them to ensure that they are getting best value from their contracts and the project.

Reasonably practicable

This term comes from the Health and Safety at Work Act 1974. It refers to the need for people to take steps to manage health and safety that are reasonably practicable – they are balanced between the potential risk of harm, and the cost and effort needed to remove the risk. The Act does not ask that all risks be removed, but that they are managed properly.

Reconciliations

A comparison of predicted or estimated against actual performance.

Rectification or defects liability period

A period of time, often six or twelve months after practical completion, within which the contractor is required to return to site to correct any defects that arise as a result of materials or workmanship not in accordance with the contract.

health, safety and environmental standards. Some clients may judge the success of their project against other criteria, such as community involvement. Some measures of project success may be more important to the client than others.

Refurbishment Construction work to an existing construction project; this may change the use or layout of the project, or simply update the materials used.

Regulations These provide the legal framework for many aspects of construction management, including health and safety and environmental management.

Resilient built environments This refers to neighbourhoods, towns and cities that are able to withstand any climate or natural forces. This may include earthquakes or floods, and the term is often used to describe future developments that will need to take climate change into account.

Resources Generally this refers to the labour, plant and materials needed to complete a task, but time and money are also key resources for the construction manager when allocating work.

Retention An amount of money held back by employers or clients from contractors. It permeates down the supply chain so that it is also held by contractors from subcontractors; and by subcontractors from sub-subcontractors. However, it is not normally held by contractors or subcontractors from suppliers. The amount is specified in contracts: perhaps 3 per cent or 5 per cent. Half of the retention would normally be released upon practical completion. The remaining half would normally be released upon the issue of a certificate of making good any defects, which is often six (or possibly twelve) months after practical completion.

Retrofit The fitting of new materials or technologies to an existing development. For example, the fitting of insulation panels to the exterior of a domestic property will be classed as retrofit works.

RIBA plan of work A commonly applied plan which sets out the sequential stages of a project for various procurement routes, and indicates when the design, information sharing and approvals for each stage of the work should be completed.

Risk Risk is the likelihood or chance of something causing harm. Often this term is used to also mean a danger or hazard, but the risk is actually the chance of this danger actually becoming an accident, also known as its probability.

Risk assessment A process by which the hazards of a task are identified and their likelihood of occurrence – risk – is estimated. The process should follow sequential steps and include the methods that will be used to minimise or eliminate the risks that have been identified. A formal risk assessment is a written document, usually on a standard template.

Risk assessment and method statement (RAMS) The two often go together, with the method statement describing the safe system of work, and the risk assessment identifying the key hazards and how they are to be eliminated, reduced or managed.

Safe system of work
As prescribed by the Health and Safety at Work Act 1974, a safe system of work is a way of carrying out a work task in the safest and healthiest way. This may be demonstrated through the use of a written method statement and risk assessment, to show that the task has been thought through and risks minimised or eliminated where possible.

Safety
The freedom from danger, risk or injury. Safety is most often linked to accidents – or, more correctly, no accidents.

Safety culture programme
A training or education programme which aims to encourage people to think positively about health and safety and to get involved in its management on sites. Often the programme is promoted by posters and within-site inductions and training sessions, to make people work in a safe and healthy way, and to encourage others to do so too.

Safety data sheets
These should accompany all materials delivered to the site. These sheets explain how the material should be managed, what PPE should be worn during its use, and how to control any accidental spill or release.

Safety management system (SMS)
A set of formal processes and procedures used to manage health and safety within a company. This may include timetabling of regular inspections, organisation-wide site rules and permits that must be used on site.

Sectional completion
This is when a project must be completed in sections, each section handed over to the client before the project has completely finished. Sectional completions are also common in housing developments, where houses may be completed in sections and sold to fund the next stage of the development.

Setting out
The practice of precisely translating construction elements, including the building itself, from the construction drawings on to the ground or other construction works. For example, the foundations will be set-out on to the ground so that the groundworkers know where they are to be dug, the walls will be set out on to the foundations so that the bricklayers know where to build, and the internal walls will be set out on to the floor slabs so that the dry-liners know where to build their partitions.

Site
The physical place where construction work is being carried out, bounded by a temporary fence or some kind of permanent boundary.

Site diary
This should be kept on a daily basis by all construction managers. The diary should record the number of operatives on site and the areas in which they were working. It should also record any key events that may have delayed work, such as bad weather, any visitors to the site, such as building control officers, and any other events of note during the day.

Site induction
Before people start work on a site, they should receive a site induction. This may be a formal or informal presentation of key information about the site to make sure people are aware of how the site is being managed and how to stay safe and healthy while working there. An induction should include information about current work activities, key health and safety risks, access routes, site rules, and other site-specific information.

Site layout
A drawing that shows the key elements of the temporary site establishment, the construction areas and all management provisions, including site offices, welfare and skips. The site layout is also a health and safety and environmental management tool, and should show all vehicle and pedestrian walkways, crossing points and emergency provision such as the muster and first aid points. The site layout should be clear and easy for anyone coming on to the site to quickly understand.

Small to medium-sized enterprises (SMEs)
SMEs can be either medium sized and employ fewer than 250 people, or small and employ fewer than 50 people, or micro and employ fewer than 10 people.

Snagging
The construction site term for the checking of any works to ensure they meet the quality requirements. Snag or defect lists are produced to list the changes and remedial work required. Snagging is often used to refer to the final completion stages of a project and carried out by contractors and subcontractors to achieve a snag-free handover to the client.

Sole traders
Individuals who are self-employed and work independently.

Specifications
Written details of the quality requirements of the construction work. For smaller projects they are usually written on the sides of drawings; on larger projects they are separate documents.

Speculative developments
These may include houses, flats or offices which are built to sell or rent as part of a development company's (which can also be a construction company) operations. The 'speculative' comes from the developer, who is speculating that they will be able to sell or rent on completion of the project.

Spill kits
These contain materials to mop up accidental chemical or fuel spills. They contain absorbent granules, pads and cotton booms to contain the spill. Spill kits should be regularly checked to make sure they are fully stocked and any used items replaced immediately.

Standard components
These are products that can be produced in vast numbers to the same specification to be used in the construction process, thereby keeping manufacturing costs down.

Strategic

Often associated with the corporate level of management, and refers to long-term thinking and planning.

Strategic Forum for Construction

The interface body between the construction industry and the government, made up of clients, contractors and workforce representatives. It aims to improve the delivery and quality of the UK construction industry

Structural engineer

The profession that focuses on the mathematic and scientific design of structures such as steel or concrete frames to ensure that they are robust and meet the structural requirements of the wider project designs.

Subcontractor

Construction work is so varied and complex that it cannot be undertaken by just one company. Therefore elements of the work are often subcontracted out to other companies, known as subcontractors. Subcontractors often specialise in a specific trade such as bricklaying, or a group of associated trades such as dry-lining and plastering, and so are subcontracted to undertake that portion of the work on a project. Subcontractors can themselves subcontract parts of the work, making sub-subcontractors.

Supervisor

Someone with a level of management responsibility for the work, often the person in charge of a particular trade on site. Supervisors may also carry out some work themselves as part of their role.

Supply chain

The linkage of companies that turn basic materials and products into the completed project for the client. Supply chains include clients, designers, contractors, subcontractors, product suppliers, materials suppliers, all the way down to the raw material extractors.

Supply chain management (SCM)

The management of all those contracted to carry out work on projects, including subcontractors and suppliers. Good supply chain management involves the development of relationships throughout the supply chain that support mutual goals and enables the development of significant competitive advantages for all those involved.

Sustainability

Sustainability is the concern for future generations. A sustainable approach aims to meet the needs of current generations in terms of resources such as water and energy, without limiting the ability for future generations to meet their own needs when the time comes. This may mean limiting our use of certain resources and seeking out alternatives that will not impact upon future generations, such as the use of solar power which is a renewable resource.

Sustainable Urban Drainage Systems (SUDS)

Drainage systems which mimic the natural movement of rainwater slowly through the ground, rather than piped systems which channel it off roofs, roads and other hardstandings which creates a flood risk when it is discharged into rivers and streams. SUDS include retention ponds, permeable paving, hardstandings and swales.

Swales Shallow grassy or vegetated dips in the ground that collect rainwater and can either channel or store it, allowing the water to slowly filter into the ground over time. Swales may be used in areas that are to remain undeveloped, and may also form part of the finished project's landscaping scheme to improve the sustainability of the project drainage system.

Team A group of people who have been assigned a goal to work towards, for example, the project team are working towards project success.

Teamwork This is the skill of working together as a team. Many teams are put together, but teamwork needs the commitment and effort of all the team members to ensure they are working as effectively and efficiently as possible.

Temporary works Any works that are needed for the construction activities to be carried out but are not part of the final construction. For example, scaffolding or other access systems are temporary works.

Tender A tender is the price submitted by the contractor to the client for which they will carry out the work. Subcontractors will also submit tenders for their specific work elements to contractors. Tender is also used to mean the processes of preparing the tender itself.

Tender price indices (TPIs) Records of historic price movements and predictions of future tender price rises or decreases that may be used alongside other data to predict future performance.

Tolerances The amount by which something can deviate from that specified and still meet the quality requirements. For example, brickwork often has a tolerance of +/-10mm, which means a wall can stray a maximum of 10mm from the setting-out line and still meet the specified quality. Tolerances should be stated within the specification for the work, and are also referred to in British Standards for construction work and materials.

Tool-box talks (TBTs) Short education and training sessions about a specific health and safety topic area delivered to a small group, such as a subcontractor's team, on site. Either held on site (around the tool-box) or in the canteen, they should be focused on a relevant aspect of work to those attending, such as de-nailing timber for a joinery gang, and the potential health and safety risks that could result from not following good practice.

Topography The way the land lies; for example, any slopes or dips or unusual features such as ponds. This can refer to both natural aspects of the land as well as man-made structures upon it.

Trades Construction work involves many different skills; one person could not become proficient in all the construction skills, and so people are trained in specific trades. For example, bricklaying, joinery and plastering are all different trades,

Traditional procurement

A procurement route where the design, construction and operation of a project are kept completely separate and are carried out independently in sequence.

Tree preservation orders (TPOs)

Written orders issued by the local planning authority that make it an offence to remove or damage a particular tree without permission. They are usually allocated because of the age, size or significance of the tree to the local area.

Two-stage tendering

A second round of tenders is often used for large construction projects. The first stage allows the client to select two or three contractors to go forward to the second stage where more details can be provided. This stops unnecessary spend on detailed tenders at the first stage of the procurement process.

Union

A representation of a collective group such as a specific industry or workforce, which can negotiate with employers over work conditions, pay and treatment.

Usability

The effective usefulness of a project once completed, ensuring that end users are able to use the building in the way it was intended.

Valuations

These are often completed at the end of each month. They are agreements between the client site and contractors' side (usually between the respective quantity surveyors) about the value of the work completed each month in monetary terms. These agreements become the basis for monthly payments, to support contractors' financing of the project.

Value engineering

A method to improve the value of a project; often thought to be a process that involves changing materials or design to give the same functionality at lower initial capital cost but which may also involve higher initial cost, with savings arising subsequently during the life cycle of the asset.

Value management

The coming together of people and processes to value engineer an element of a building.

Waste

Frequently means any materials that are not used in the construction of the project, and end up being thrown into skips. Waste can be caused by damage to materials, poor storage, poor site management or poor design, resulting in many offcuts which cannot be reused. Waste can also refer to waste of time and efforts. For example, a change to the location of a wall on site can result in wasted time and efforts for the bricklayers, wasted bricks in the demolished wall that cannot be reused, and wasted money that has to pay for the labour time and new materials.

and the tradesmen, or operatives, all need different skills to be able to carry out their work correctly. Trades may also be termed 'wet' – where water is involved in their work such as plastering – or 'dry', where there is no water involved such as joinery.

Waste hierarchy *This is set out by the Waste Regulations 2011, which state an order of preference for the treatment of waste materials produced on site. The hierarchy from the most preferable method follows prevention, reuse, recycling, other recovery and disposal. Application of the hierarchy must be made for all waste leaving site, and recorded on waste transfer notes.*

Water run-off *This is water that either falls as rain or is produced through industry activities such as spraying or damping down, and then moves into the drainage systems or watercourses. Water run-off will collect any debris in the form of silt or mud, and may pick up chemicals and other pollutants through contact with materials, waste or contaminated ground. Water run-off should be controlled and managed in order to reduce this pollution pathway.*

Welfare *Used to describe what should be provided for the welfare of the workforce; this includes toilets, washing facilities, drinking water, changing rooms and lockers, drying rooms for wet clothes, and places to prepare food, eat and rest. Welfare requirements are legislated for through the CDM Regulations.*

Well-being *This is becoming more prominent within construction management, as a workforce with good levels of well-being is likely to be more effective and efficient in their work. Well-being not only means being safe and healthy but also not tired, stressed or unhappy at work.*

Wheel washes *These can be installed at the site gates to clean the wheels of any wagons or delivery vehicles that have come on to the site and are now leaving with muddy wheels. Mud tracked on to the local roads can be a skid risk, it can dry and turn to dust and it can also form pollution as it is washed into the surface water drains. Wheel washes should contain the water they use and recycle it if possible to conserve resources and prevent muddy water run-off.*

Whole life cycle *The construction life cycle starts from client need and progresses through the design, construction, operation and maintenance of a project until it is demolished or refurbished by the client.*

Work packages *These are used to bundle elements of construction work together. This may be by trade or part of the project, for example, the brickwork package or the envelope works.*

Working rule agreements *These are national agreements that have been produced between the unions and employer organisations to set standards for working conditions and pay.*

WRAP *The government-funded body focused on reducing waste and using resources sustainably. It produces guidance about waste reduction for construction industry clients, designers and contractors.*

Zero carbon   *This does not yet have a definitive definition. It can mean that a development will not generate an overall carbon footprint, but that the carbon used in the materials and construction will be offset against carbon that is not used in its operation and maintenance. For example, if a development can generate all its own electricity, this will be zero carbon in terms of energy use, but in practice this is very hard to achieve.*

Suggested exercise solutions

Exercise A

Consider the remaining principles as they are detailed in Table 2.3.2. For each principle, consider the scenario and think about the following:

- What could go wrong if work was not carried out following the principle?
- How could construction managers implement the principle?

Please see Table 2.3.3. for suggested answer.

Exercise B

You are working as the construction manager for a main contractor, and under your current design-and-build contract you are involved with managing the design team. While the architect and engineer seem happy to work together and share information, they haven't always included the mechanical and electrical (M&E) services designer. This has already led to one costly redesign of the plant room.

- What problems could have arisen here?
- In your role as the manager, how would you try to solve them?

There seems to be a communication issue here.

This could be caused by the way design information is being shared technically – perhaps the architect and engineer use the same cloud computing system, making it easy for them to quickly share information, but the M&E designer does not and so has been left out of the communication loop. If this is the case, then the manager should ensure that everyone is using the same system and can easily share information to avoid delays to one member of the team.

This may also be due to a lack of understanding in the receipt of communications on the part of the M&E designer: does everyone know which drawings are for information and which contain design changes

Table 2.3.3 Exercise A suggested answer

Management Principle	What could go wrong if this was not implemented?	How could construction management implement this principle?
Authority and responsibility – Managers must be able to give orders, which means they are also responsible for them.	If no management decisions are made, and no orders given, time can be wasted as the workforce try to come to a decision among themselves – and they may not want to do this. But if the manager does not have authority and their decisions are ignored, the workforce may follow their own ideas which could lead again to wasted time and money if these do not fit with other plans. If managers do not take responsibility for their orders, even if they are wrong, their workforce can lose motivation. If the manager tries to blame someone else this can cause more bad feeling and arguments. The manager is not likely to gain support from others in the future and will lose their authority on the site.	The site manager must plan the work and clearly instruct the gangs as to the sequence of the houses and the time allocated for each aspect of the work. They should call in the concrete which will provide the end milestone for the task – if this is early or late they must then take responsibility for the repercussions of that, and not pass on blame to the site operatives.
Discipline – Employees must follow the rules, but there should be careful use of punishment.	Discipline is important to ensure that people stay safe and are working effectively and efficiently – but this can be abused. If managers become tyrannical and start to punish people without real reason, there will be bad feeling and lack of motivation on the site, affecting production.	The site manager should make sure everyone is following the site rules, including the health and safety rules. They may need to discipline those who do not follow the rules; this may delay the work but the manager is responsible for health and safety on the project and so should take the necessary steps to accept the delay for a safe and healthy site. Discipline should only be used when needed.

Unity of command – Each employee should only receive orders from one superior.	If the workforce are getting different orders from different people then which is the right course of action becomes very unclear. This can result in delays of time while people check and double check what they are to do, or they carry on and could produce incorrect work that has to be taken down. This is a waste of time and money and again can de-motivate the workforce – especially if they are blamed for the miscommunication.	The site manager may give instructions to the foreman of the groundworks gang. The foremen will then pass on the information to their individual operatives. This means that the site manager is not trying to speak to too many people at once, and each operative has a clear message from their immediate superior. If too many people are giving orders they can become confused, and can lead to a lack of coordination which could cause health and safety risks; for example, if work shifted out of sequence due to miscommunication.
Unity of direction – One manager in control of all the activities leading to one final objective.	If a different manager is in charge of each house foundation there may be conflict, as they would each want to complete their work first, rather than looking at the bigger picture. If activities are completed out of sequence this can cause problems for the overall programme.	One construction manager should be in control of all 20 houses and all the tasks leading to their completion, including the casting of the 20 foundations.
Subordination of individual interests to the general interest – Everyone should be working together towards the common final objective, not their own individual goals.	If each member of the groundworks team is only working for their own benefit (for example, to complete their work the fastest) this could have a negative effect on those working on tasks behind or ahead in the sequence. This could lead to standing time or work shifting out of sequence and becoming inefficient.	Although the different trades may only be keen to complete their work (for example, the excavator driver only wants to complete the excavation) and not the final objective which would be the completion of the foundations, they must still become part of the team to meet the final objective. This principle can be especially problematic for site managers on larger projects where many different subcontractors must work together to complete the project, but are often focused on just their own element of the work.
Remuneration – A fair day's pay for a fair day's work.	If pay is not fair the workforce will not stay. There could be disputes, bad feeling and even strikes.	The site manager must ensure that the pay meets any national agreements and is fair, and any extras such as wet weather working are allowed for.

(Continued)

Table 2.3.3 Continued

Management Principle	What could go wrong if this was not implemented?	How could construction management implement this principle?
Centralisation – Decisions about work can be kept centralised with the manager, or decentralised to others; a balance is needed for the right decisions to be made by the right people.	If the manager has to make every little decision, they will become preoccupied with unnecessary details and not have time to look at the bigger picture. This can lead to mismanagement in other areas; for example, too much focus on the foundations may mean a lack of planning and preparation for the brickwork and roofs of the houses, and could cause problems in the next few weeks as new trades come on to site.	While the site manager should make the key decisions about the project – sequence and timing, etc. – the foremen can make other decisions, for example, where best to site the reinforcement materials so that they can be easily and safely moved on to the plots as needed.
Scalar chain – The management chain. The level of authority reduces down the scalar chain, and there must be clear communication in both directions. It can be broken if this is in the interests of the business.	Without a clear scalar chain the authority of the site will not be established. If there is no clear manager, decisions can become confused and an overall direction for the project can become lost. Lack of communication can mean that problems are not raised, and can grow into serious issues, for example, problems with workforce morale.	The scalar chain shows where the site manager sits within the lines of authority and communication. This supports their authority on the site and makes sure they are able to maintain unity of command and direction, as well as discipline as needed. It also ensures clarity in communication. The site manager should make sure their subordinates in the chain, such as the foremen, are involved in all communications to the operatives, to make sure everyone is aware of the plan, sequence or health and safety risks of the work. The chain should be broken if it is in the best interests of the company. This could include health and safety concerns raised by the workforce. If these are not acted upon by the immediate link above in the chain, the chain should be broken and the workforce raise their concerns higher up the chain.

Order – A place for everything, and everything in its place, including labour, plant and materials.	Groundwork can be very difficult to control in terms of site layout and materials as the work is being carried out in the ground itself. Poor materials management can result in unsafe sites, with risks of trips or falls, and wasted materials as they sink into the mud. If members of the groundworks team keep moving roles, they will not be keeping to the division of work principle and so there can be drops in quality and efficiency.	It is vital that clear stores are established for materials and that they are kept tidy through good housekeeping to eliminate health and safety risks of trips or falls. Plant should also be kept in designated areas to ensure that it is well cared for. Each member of the groundworks team must stay in their role – they should not switch around without planning, as that is also when mistakes and accidents can happen.
Equity – Managers must be fair and just to their employees.	If managers are not fair there can be serious repercussions, people can leave or they can become very upset and angry at the management. An unhappy workforce is de-motivated and unproductive.	Site managers must be fair to the site operatives. If they have been working through bad weather and the manager has promised them an early finish they must deliver, and not suddenly find another job to be completed before they leave.
Stability – *of personnel* – Low rates of labour turnover mean more efficient production.	For this size of project, the team should remain the same throughout. Problems can arise with changes in personnel; for example, the groundworkers will have become used to communicating with the excavator driver, and if they should change then new communication methods may be needed which can delay production, or indeed cause new health and safety risks.	The construction manager should ensure that the site team remains the same, so that people are able to develop good working relationships and become more efficient as a team. In the construction industry this can be hard to maintain, as different trades and skills can come and go on a project.

(Continued)

Table 2.3.3 Continued

Management Principle	What could go wrong if this was not implemented?	How could construction management implement this principle?
Initiative – Employees should be able to bring new ideas to their work.	If they are not allowed to have input into their work, some employees become disheartened. People are likely to be more engaged and motivated to work if they are able to become part of the bigger picture. Treating the workforce as just labour can also mean that managers miss out on new ideas and better ways of doing things.	There may be scope for new ideas on the project, and the site manager should listen, consider and discuss them with their team. For example, if the ground is proving to be poorer than first thought, a groundwork operative may suggest the use of a pre-form system such as pecafil, which would be faster than traditional shuttering and bring the work back on programme, albeit at a cost.
Esprit de corps – Team spirit!	Without team spirit, a construction site can be a very negative place. If people are only out for themselves, conflicts and arguments are common and the site becomes an unpleasant place to work in – people may leave or become de-motivated, with reductions in production and quality.	Team spirit means a happy workforce who are keen to work together to achieve the final objective. This can also improve health and safety on site, as people look out for each other more, and are more willing to speak up about potential problems and hazards.

to be made to other work? The manager may need to establish a system to clarify that changes need to be made – possibly an electronic system for comment on the drawings as they are issued to the team.

Another consideration is whether there is a clash of personalities that is affecting the communication flow and the M&E designer is being deliberately left out. In this case, the manager may need to take charge and firmly but fairly raise the issue during a meeting with all design team members present and clearly remind the team that failure in any part of the project reflects badly upon all team members, and that working together is vital for success.

There may also be other possibilities too.

Exercise C

You are a construction manager and have been tasked with managing the joinery package on site. However, one of your joinery gangs is not meeting the required outputs in fixing the doors on the first floor, and you can see they are falling behind programme. When you speak to the gang they tell you they are making enough money with the bonus rate they are being paid, and there isn't any need to go faster. However, if they don't meet the programme this will have a knock-on effect on the following decoration trades.

- What motivational factors are in play in the above scenario?
- How could you resolve the issue to stop this affecting the programme?

In this scenario, money has been used as the motivator. But this has not been implemented correctly – the bonus payment is too large compared to the work rate and so the joiners have slowed down their output as they are happy with the money they are making at this rate.

The joiners do not seem to be only motivated by money as they could be earning more already if they wanted to, so to increase the bonus to speed up the work is unlikely to have any effect. The opposite option is to reduce the bonus to make the joiners speed up, but that could prove very unpopular and may even lead to the joiners leaving the site.

The best approach would be for the construction manager to try to communicate with the joinery gang, and clearly explain the problem their rate of production is causing to the overall programme. If they are not rational, economic men they may be self-actualising men who would relish the challenge of bringing the work back on programme. Or they may be social men, and happy to help out the wider project team.

The construction manager needs to appeal to the most suitable motivator for this gang, and in future ensure that bonus rates are recalculated for the next joinery tasks to the correct levels. Remember the management principles: if it is the manager's decision to set these rates then they are responsible for the repercussions, and resolving any problems that may arise from their decisions.

Exercise D

You have just been transferred to a new domestic development site to take over from another construction manager who has left the company. While familiarising yourself with the site, you meet the brickwork foreman who seems to be in charge of most of the works on site! While you are talking to him, several other trades come up to ask for information or advice, and he seems happy to help. However, you notice that some of his ideas might help complete the brickwork quickly and efficiently, rather than the houses and the project as a whole.

- How could this shift in leadership from the formal management team have occurred?
- What problems could this cause you in the future, and how could you solve them?

This foreman has emerged as the informal leader of the site and he seems happy to take on the role, as it enables him to complete his work as effectively and efficiently as possible. The previous site manager may not have been a strong enough character to stop this informal 'take-over' or he may not have developed as much rapport with the site workforce. Alternatively, if he wasn't a good construction manager, then he may not have had the answers and information needed to help the other trades, resulting in a shift in authority to a more competent foreman instead.

Problems will arise when the goals of the brickwork foreman start to diverge from those of the project team. He will be concerned with completing his allocated portion of the work, and will not be considering the follow-on trades and their requirements. This may cause problems if work is completed out of sequence or simply to the benefit of the brickwork – all trades must work together to complete the project successfully.

The best way to solve this problem is to work alongside the foreman to regain control of the project; any direct hostility will only cause problems and arguments. By demonstrating the necessary knowledge and skills of a good construction manager, and making sure people know who you are and to come to you for information, you can shift leadership back to the project team where it should be. Without unity of direction (one of Fayol's management principles) the project will be all about the brickwork and not about the houses.

It may be a good idea to revise the project programme and issue it personally to all the other trade teams and foremen. This would make sure that they are aware of who you are, that you have the overall plan and are in charge of the project. While the bricklaying foreman may be a little put out, he will also likely be relieved that he doesn't have to be disturbed so often, and that you are now there to help him out with information as needed.

Exercise E

You are the construction manager on a new-build domestic site. You notice that the new junior quantity surveyor is not pulling his weight and seems to spend most of the day outside in the smoking hut talking to the operatives. When you tackle the senior quantity surveyor he seems reluctant to talk about it and doesn't appear to be very concerned.

- What aspects of group theory could be happening here?
- How could you bring the junior QS back into the project team?

The junior QS does not seem to want to be part of the project team, and feels more comfortable with the operative team on the site. He may have started out on the tools and therefore feel that he is still part of that team and be comfortable in that environment, rather than with his new team in the office. Or he may simply have found the operative group more welcoming and, feeling out of place in the office, has decided to spend his time in the smoking hut instead.

This lack of team integration could be caused by the attitude of the senior QS, who does not seem bothered about the situation and may be excluding the junior QS from his team. But there are management issues here, as work is not being carried out efficiently if one member of the commercial team is never in the office. There could be personality issues that have meant the QS team has not gelled. Alternatively, the senior QS simply may not be interested in having a new team member in his department. But part of any senior role is the management and support of more junior members of the team, and so a more frank talk may be needed with the senior QS to remind him of his responsibilities and encourage him to support and help the new QS develop.

As the construction manager, the first step would be to try to bring the junior QS into the wider project team, and encourage him to focus on his new professional role; you could do this by offering to become their mentor, to help them 'find their feet'. Hopefully the senior QS would realise that they do need to take action, and a change in their attitude may spark a change in the junior QS's attitude too. In the worst case scenario, the junior QS is not interested in his new role, and so if his attitude remains when others have changed, and if he is not willing to become a team player, other action may need to be taken.

Exercise F

- What activities need to be added to the above list to complete the house?
- Assuming a two-storey detached construction, list the activities to help the future development of weekly programmes.

The activities for the rest of the house could include the following:

- Scaffolding.
- First-floor joists.
- Brick and block to roof.
- Wall plate.
- Roof trusses (fit and brace).
- Soffits and fascia.
- Roof coverings (felt, batten and tile).
- Guttering.
- Windows.
- External doors.
- First fix joinery (stairs, first floor, internal walls one side).
- First fix plumbing (pipe work within floor voids).
- First fix electrical (conduits and wiring in walls).
- Second fix joinery (second side internal walls and ceilings).
- Plasterwork.
- Decoration mist coat.
- Final fix joinery (skirting, internal doors).
- Final fix electrical (light switches, socket faces).
- Final fix plumbing (radiators, ceramics, taps).
- Decoration final coats.
- Kitchen fit-out.
- Bathroom fit-out.
- Floor finishes.
- Testing.
- Snagging.
- Final clean.
- Handover.

Exercise G

An excavator and driver have a performance output of 25m³ per hour (constant): 10,000 cubic metres of spoil need to be excavated to prepare the sites.

- How long will this take?

Machine performance output = 25m³/hr
Amount to remove = 10,000m³

$$\text{Duration} = \frac{10{,}000\text{m}^3}{25\text{m}^3/\text{hr}} = 400 \text{ machine hours}$$

$$\text{Working time} = \frac{400\text{hrs}}{38\text{hrs/week}} = 8.51 \text{ weeks}$$

Exercise H

A wall needs to be built to the front of a domestic property garden. The specification is as follows:

- 1.8m high.
- Length of the wall will be 25m.
- Coping to the full length.

Assume there are 59 bricks per metre squared and the bricklayer can lay 50 bricks per hour.

The coping stones are 600mm long and it takes the bricklayer 10 minutes to lay one coping stone.

- How long will it take one bricklayer to complete the wall?

Size of wall = 1.8m x 25m = 45m²

Amount of bricks in wall = 45m² x 59 bricks/m² = 2 655No
Rate to lay = 50/hr

$$\text{Duration to lay bricks} = \frac{2\,655\text{No}}{50\text{bricks/hr}} = 53\text{hrs}$$

$$\text{Number of coping stones} = \frac{25\,000\text{mm}}{600\text{mm}} = 41.6\text{No}$$

Duration to lay copings = 41.6No × 10mins / stone

$$= 416\text{mins} = \frac{416}{60} = 6.9\text{hrs} = 7\text{hrs}$$

Total hours' work = 53hrs + 7hrs = 60.1hrs = 1.58wks

For one bricklayer this would mean 1.58 weeks' work. However, bricklayers work in gangs and so this rate needs to be adjusted accordingly. Often two bricklayers will work with one labourer (known as a 2+1 gang) and so the rate may be adjusted to reflect this:

$$\text{Duration of work} = \frac{1.58\text{wks}}{2} = 0.79\text{wks}' \text{ work or}$$

approximately four days for the 2+1 gang.

Exercise J

It takes a floor layer 0.46hrs to lay 1m² of flooring.

- How long will it take one floor layer to complete the first floor which is a total of 425m²?

$$\text{One fitter can lay } \frac{1}{0.46} = 2.17\text{m}^2/\text{hr}$$

$$\text{Duration} = \frac{425\text{m}^2}{2.17\text{m}^2/\text{hr}} = 195\text{hrs} = 5.15\text{wks}$$

Again, if two floor layers were working together the duration would be reduced as follows:

$$\text{Duration of work} = \frac{5.15\text{wks}}{2} = 2.57\text{wks' work or}$$

approximately 13 days for the two floor layers.

Exercise K

Consider Figure 3.3.5: this shows an internal disabled toilet fit-out. Think about the space as a 'shell' – the blockwork walls, concrete soffit ceiling and concrete floor slab. Out of picture are a hand-drier, suspended ceiling and emergency alarm.

Carry out the planning for the fit-out of the shell to the level of finish shown in Figure 3.3.5, at the level of detail needed to produce weekly construction programmes for the work:

- List the activities needed to fit out the toilet.
- Place these activities in the optimum sequence for the work to meet the technical requirements and minimise damage to existing work.
- List the trades that would carry out each activity, and decide if everything needs to be done by each trade in just one visit.

Table 3.3.1 shows the key activities, the trade and the sequence needed to carry out the work.

Table 3.3.1 Exercise K suggested answer: sequencing work

Activity	Trade	Sequence
Mist coat to walls (first coat)	Painters	1
First fix mechanical – pipe work and fixings	Plumbers	2
First fix electrical – conduit	Electricians	3
First fix alarm – conduit	Alarm fitters	4

Table 3.3.1 Continued

Activity	Trade	Sequence
First coat of paint	Painters	5
Suspended ceiling grid with patressess tiles (single tiles with plywood sat behind to provide a fixing for light fitting and alarm)	Ceiling fixers	6
Vinyl flooring and sit-on skirting (and protection once complete)	Floor fitters	7
Second fix mechanical – ceramics and fittings, complete pipe work	Plumbers	8
Second fix electrical – pull cables, light fitting and hand-drier	Electricians	9
Second fix alarm – pull cable and fix in ceiling	Alarm fitters	10
Final coat of paint	Painters	11
Final fix mechanical – fill and test system	Plumbers	12
Final fix electrical – final connections, testing and commissioning	Electricians	13
Final fix alarm – final connections, testing and commissioning	Alarm fitters	14
Fit ceiling tiles	Ceiling fixers	15
Final coat of paint	Painters	16
Fittings – disabled handles, soap and towel dispensers	Joiners	17
Final clean	Cleaners	18

Exercise L

Activity	Duration	Date 1	2	3	4	5	6	7	8	9	10	11	12	13	14	15
Start	0 days															
Activity A	2 days															
Activity B	6 days															
Activity C	3 days															
Activity D	4 days															
Activity E	3 days															
Activity F	1 day															
Completion	0 days															

Exercise M

Activity	Duration	Date														
		1	2	3	4	5	6	7	8	9	10	11	12	13	14	15
First side plasterboard	6 days															
First fix electrical	9 days															
Second side plasterboard	3 days															

Exercise N

Activity	Duration	Date														
		1	2	3	4	5	6	7	8	9	10	11	12	13	14	15
Start	0 days															
Activity A	2 days															
Activity B	6 days															
Activity C	3 days															
Activity D	4 days															
Activity E	3 days															
Activity F	1 day															
Completion	0 days															

Exercise P

Activity	Duration	Date														
		1	2	3	4	5	6	7	8	9	10	11	12	13	14	15
First side plasterboard	3 days															
First fix electrical	6 days															
Second side plasterboard	3 days															

Exercise Q

Following the HSE's five steps to risk assessment:

1 Look for the hazards:

• machinery operating on the site;
• uneven ground;

- excavations;
- material stores.

Barriers have been used to manage these hazards.

The work areas around the excavations, the materials store and the machines have been segregated from the walkways using pedestrian crash barriers.

While these appear to be in good order, they can be easily moved and damaged within this kind of working environment and a more robust barrier may be needed when the site becomes more congested.

2   Who might be harmed and how (e.g. machine drivers going to and from their machines, other site workers)? The public are segregated by the hoarding shown in the distance. Those at risk could be struck by a machine, trip or fall on the uneven ground, or fall into an excavation.

3   Evaluate the risks: at present the barriers have segregated people from the work areas clearly. From the ESCAPE hierarchy the risks cannot be eliminated or substituted, but they have been controlled.

4   Record the findings: this should be on a risk assessment form.

5   Review and update: when the next phase of the groundworks begins, possibly drainage or concrete foundations, the works must be reassessed to ensure that safe systems of work and a safe working area are maintained.

Programme of work																		
			Date															
Activity	Duration	Resources	1	2	3	4	5	6	7	8	9	10	11	12	13	14	15	

Blank risk assessment template

RISK ASSESSMENT

PROJECT:									
TASK:									

	Hazards		Who Might be harmed and how?	Evaluation		Precautions to be Taken		Action By	Residual Risk
				Hazard	Risk				
1.									
2.									
3.									
4.									
5.									
Assessed By:		Date				Approved By:		Date	

Note: This RA is to be reviewed if there are any changes of work method, changes to the work environment or deviation from the method statement.

Review 1:		Date				Review 2:		Date	

Index